AF324740

WERNER BARTSCH

DESERT BIRDS

WERNER BARTSCH

KEHRER

für Tim & Nina

A cockpit faces into an endless desert landscape; a gangway stands in the middle of an abandoned expanse, points at the sky and appears to wait for planes that no longer fly, for passengers that no longer arrive. In this kilometer-wide desert in the southwestern USA, hundreds of unused and discarded aircraft stand in waiting and contemplate their fate. While one normally encounters them amidst the hustle of the airports of this world, as the center of a network of activities, here the planes are found in absolute solitude. Not a single person far and wide, no infrastructure, no movement. It is not artificial materials like cement, tar, steel, or glass that make up the scenery, but instead sand, a few scattered grasses, and rocks.

This isolated and deserted location appears absurd and bizarre for the once so modern and freedom-connoting airplanes. So many people taken to distant places. So many dreams and longings, fears and joys embodied. So many people who trusted them with their lives and hopes. Countless stories and memories slumber in these ships of the air and are now fading away into the remote distance. The only reminders of the activities of the past that remain are the imprints and furrows in the hot desert sand.

Whether these aircraft will rise into the air one more time like a phoenix from the ashes is uncertain. Several of them have already stood here for decades and are aging quietly. What begins as a stopover in times of crisis and as temporary disuse not infrequently becomes the last stop. Obsolete or defective, the planes remain useful only as a stockpile of replacement parts. These parts are removed and resold; the unusable remnants are disposed of and melted down.

Deprived of their mobility and functionality, the discarded planes are reduced to their basic form. In the gleaming light of the burning desert sun, on white sand and against a backdrop of bright-blue sky, the aerodynamic shapes and elegant curves shine with an unreal brilliance. In the sleek wings and abstract silhouettes, which stand out in the bright glow of the moon and under the clear, starry sky, the futuristic expressiveness of these aircraft becomes apparent one more time.

In his photography, Werner Bartsch concentrates on the "new" that comes about when familiar relationships between subjects and surroundings are dissolved. His works show the contradiction between the subject aircraft and the desert context as well as the tension that this contradiction produces. The vertical line of focus in some pictures draws the eye to details and structures, to individual planes in endless rows, to gleaming surfaces, and to a plane's ragged fuselage, from which two lonely seats look out into the open. In its expansiveness and simplicity, the arid desert functions as a minimalist backdrop against which the grace and beauty of the "Desert Birds" are made fully evident. Like imposing metal sculptures, they submit to their rough surroundings, becoming unintentional works of art placed in no-man's-land.

The very specific interplay of colors and forms, the mood of the lighting, and the endlessness of the barren surfaces combine to form a unique aesthetic through which the fascination that these places and subjects generate becomes perceptible. These photographs are not inventories of discarded airplane models, not documentaries of storage yards and airplane graveyards; they are a tribute to countless hours in the air and long journeys, to the pioneer spirit and human imagination.

Sophia Greiff

Ein Cockpit weist in eine endlose Wüstenlandschaft; eine Gangway steht inmitten einer verlassenen Weite, zeigt gen Himmel und scheint auf Flugzeuge zu warten, die nicht mehr fliegen, auf Passagiere, die nicht mehr kommen. In den kilometerweiten Wüsten im Südwesten der USA stehen Hunderte von stillgelegten und ausrangierten Flugmaschinen auf Halde und blicken ihrem Schicksal entgegen. Während man ihnen sonst stets in der Betriebsamkeit der Flughäfen dieser Welt begegnet, als Zentren in einem Geflecht von Aktivitäten, trifft man sie hier in der absoluten Einsamkeit. Kein Mensch weit und breit, keine Infrastruktur, keine Bewegung. Nicht künstliche Materialien wie Beton, Teer, Stahl oder Glas prägen die Szenerie, stattdessen Sand, ein paar vereinzelte Gräser und Steine.

Absurd und bizarr wirkt dieser abgeschiedene und menschenleere Standort für die einst so modernen und Freiheit verheißenden Luftfahrzeuge. Wie viele Personen haben sie an entfernte Orte gebracht. Wie viele Träume und Sehnsüchte, Ängste und Freuden haben sie verkörpert. Wie viele Menschen haben ihnen ihr Leben und ihre Hoffnungen anvertraut. Unzählige Geschichten und Erinnerungen schlummern in diesen Luftschiffen und verhallen nun in den abgelegenen Weiten. Als Reminiszenz an die vergangene Aktivität bleiben nur die Abdrücke und Furchen im heißen Wüstensand.

Ob diese Flugzeuge noch einmal wie ein Phönix aus der Asche in die Lüfte emporsteigen werden, ist ungewiss. Einige von ihnen stehen bereits seit Jahrzehnten hier und altern leise vor sich hin. Aus einer Zwischenstation in Krisenzeiten und der vorübergehenden Nichtverwendung wird nicht selten eine Endstation. Veraltet oder defekt sind die Maschinen nur noch als Ersatzteillager von Nutzen. Bauteile werden ausgebaut und weiterverkauft, unbrauchbare Überreste werden entsorgt und eingeschmolzen.

Ihrer Mobilität und Funktionalität beraubt, sind die abgestellten Fluggeräte reduziert auf ihre reine Gestalt. Im gleißenden Licht der brennenden Wüstensonne, auf weißem Sand und vor leuchtend blauem Himmel, erstrahlen die aerodynamischen Formen und eleganten Rundungen in unwirklichem Glanz. In den schnittigen Tragflächen und den abstrakten Silhouetten, die sich im hellen Schein des Vollmondes vor klarem Sternenhimmel abzeichnen, wird noch einmal die futuristische Ausdruckskraft der Flugzeuge spürbar.

Werner Bartsch konzentriert sich in seinen Fotografien auf das Neue, das entsteht, wenn gewohnte Beziehungen von Objekt und Umgebung aufgelöst werden. Seine Arbeiten zeigen die Widersprüchlichkeit zwischen dem Objekt Flugzeug und dem Kontext der Wüste sowie die Spannung, die sich aus diesem Widerspruch ergibt. Die in einigen Bildern vertikal verlaufende Schärfenebene lenkt den Blick auf Details und Strukturen, auf einzelne Flieger in endlosen Reihen, auf glänzende Oberflächen und einen abgerissenen Flugzeugrumpf, aus dem zwei einsame Sitzplätze ins Freie blicken. Die ausgedörrte Wüste in ihrer Weite und Schlichtheit fungiert als reduzierte Kulisse, in der die Anmut und Schönheit der „Desert Birds" zur vollen Geltung kommen. Wie imposante Metallskulpturen fügen sie sich in ihre unwegsame Umgebung, werden zu unbeabsichtigten Kunstwerken, aufgestellt im Niemandsland.

Das ganz spezifische Zusammenspiel von Farben und Formen, die Stimmung des Lichts und die Endlosigkeit der kargen Flächen ergänzen sich zu einer einzigartigen Ästhetik, in der die Faszination, die von diesen Orten und Objekten ausgeht, spürbar wird. Diese Fotografien sind keine Bestandsaufnahme ausrangierter Fliegertypen, keine Dokumentation von Lagerplätzen und Flugzeugfriedhöfen – sie sind eine Hommage an unzählige Stunden in den Lüften und weite Reisen, an Pioniergeist und menschliche Imagination.

Sophia Greiff

It's late on Sunday afternoon as I enter the four-digit access code into the little box on the gate. Dark-blue sky. Crystal-clear light. The air smells of dust and old oil. The gate opens and grants me passage onto the airfield. It's no longer so hot – the sun's rays stretch long over the desert – and I can sense the nearing cold of the night. I drive on crunching sand. In the distance, bizarre mountain ranges. In front of them, a nothingness of sand and shrubs.

The object of my desire has stood here for several years. Already from a distance it gleams silver like a piece of art. This is my second visit to the VC-121A Constellation, once President Eisenhower's Air Force One, with the serial number 48-610. It looks well maintained. Some of the hatch covers on the engines hang towards the ground, the rudders are missing from the triple vertical tail, but otherwise it's in good shape. The sky above it is like the backdrop in a perfectly directed theater piece. Everything is right. Including the quiet.

I put my camera on the tripod, not only aware that I'm about to photograph one of the most artful designs of flying history in an unbelievable setting, but also with a certain reverence for the story of this aircraft, built in 1948 in Burbank, California. In 1953 it became the presidential airplane and was named Columbine II after the official flower of the state of Colorado, Mrs. Eisenhower's home state. Following its last presidential assignment at the end of 1959, it flew for several more years as a government's VIP plane, going into retirement in 1968 with 14,072 flying hours. Its fate since then is unclear.

I enjoy the solitude, shoot a few extra times simply because of the moment. I observe the light and the changes in shape. Not a soul far and wide. The impression of solitariness is altered abruptly as I get the feeling that I am being watched. I notice a four-legged being coming towards me out of the by now diffuse nothingness. Alarm bells ring inside me. For a city dweller who knows dangerous creatures only from TV documentaries, this is a true challenge. My biological perception assesses the animal as, one, too big; two, too wild; and three, underway without a master

or mistress. I label it some kind of husky – a sled dog. But in the desert? I quickly decide (in my thoughts I'm already battling the wild beast in the desert sand) on a stopover in my car, which is parked about 30 meters away and offers the quickest option for getting out of the danger zone. I simultaneously remember those same TV documentaries, in which fleeing people were always caught by the angry animals and thrown to the ground. So I start my retreat with small steps, the quadruped always in sight, ready to run in a panic if there's an emergency. When I'm finally sitting out of danger in the passenger seat, I feel completely stupid.

I see my camera, the light, the Constellation, the desert. My photographer's heart burns. The light fades, and the Constellation's gracefulness transforms almost by the second. My four-legged companion is now close to my camera; he sniffs and pees in the desert sand. The mood stays peaceful – and so does he.

I get out again and go to my camera to keep working. I try to ignore the danger and realize that it never existed. My canine observer never comes closer to me than 10 or 15 meters. He sniffs, watches me on and off as I work and adjust the lighting, pees in the sand again – at one point even on "my" Constellation. We have, so to speak, a peaceful coexistence, which even turns into a reciprocal liking – at least on my part.

I continue photographing the scenery until no more light exists and it's cold and dark. My four-legged companion has simply disappeared – unfortunately I never photographed him. Days later I tell my story to an employee at the regional airport. There are no huskies in the desert, he says, but definitely coyotes interested in photography. I will never forget this meeting of animal, human, and machine.

Werner Bartsch

Es war später Sonntagnachmittag, als ich den vierstelligen Zugangscode in das Kästchen am Gate eintippte. Dunkelblauer Himmel, kristallklares Licht, es riecht nach Staub und Altöl, das Gate öffnet sich und gibt die Durchfahrt auf das Airfield frei. Es ist nicht mehr so heiß, die Strahlen der Sonne ziehen sich lang über die Weite, und man erahnt die nahe Kälte der Nacht. Ich fahre auf knirschendem Staub, in der Ferne bizarre Gebirgsketten, davor ein Nichts aus Sand und Sträuchern.

Das Objekt meiner Begierde steht hier schon seit einigen Jahren. Silbrig glänzend wirkt sie schon aus der Ferne wie ein Kunstwerk. Es ist mein zweiter Besuch bei der VC-121A Constellation, der früheren Air Force One von Präsident Eisenhower, mit der Kennung 48-610. Sie sieht gut erhalten aus, einige Klappen an den Triebwerken hängen nach unten, am Dreifachleitwerk fehlen die Seitenruder – aber ansonsten ist sie in good conditions. Der Himmel über ihr wirkt wie eine Kulisse in einem perfekt inszenierten Theaterstück. Alles stimmt. Auch die Stille.

Ich stelle meine Kamera auf das Stativ – nicht nur mit dem Bewusstsein, eine der kunstvollsten Formen der Flugzeuggeschichte in einer unglaublichen Umgebung zu fotografieren, sondern auch mit einer gewissen Ehrfurcht vor der Geschichte dieses Flugzeugs, erbaut im Jahre 1948 in Burbank, Kalifornien. 1953 wurde sie zur Präsidentenmaschine und wurde Columbine II genannt, nach der offiziellen Blume des Staates Colorado, Heimatstaat von Mrs. Eisenhower. Nach ihrem letzten präsidentialen Einsatz Ende 1959 flog sie noch einige Jahre als VIP-Maschine der Regierung und ging mit 14.072 Flugstunden 1968 in Rente. Seitdem ist ihr Schicksal ungewiss.

Ich genieße die Einsamkeit, betätige den Auslöser einige Male öfter, nur um des Momentes willen. Ich beobachte das Licht und die Veränderung der Form. Keine Seele weit und breit. Der Eindruck der Einsamkeit ändert sich schlagartig, als ich das Gefühl habe, beobachtet zu werden. Ich bemerke ein vierbeiniges Wesen, das aus dem mittlerweile diffusen Nichts gemächlich auf mich zukommt. In mir klingeln die Alarmglocken. Als Städter, der gefährliches Getier nur aus Fernsehreportagen kennt, eine wahre Heraus-

forderung. Meine biologische Wahrnehmung stuft das Tier als erstens zu groß, zweitens zu wild und drittens ohne erkennbares Herrchen oder Frauchen unterwegs ein. Ich kategorisiere eine Art Husky – einen Schlittenhund. Aber in der Wüste? Kurzum, ich beschließe (in Gedanken schon mit der wilden Bestie im Wüstensand kämpfend) einen Zwischenaufenthalt in meinem Wagen, der circa 30 m entfernt steht und mir die schnellste Möglichkeit bietet, aus der Gefahrenzone zu kommen. Gleichzeitig erinnere ich mich an die gleichen Fernsehreportagen, bei denen flüchtende Personen stets von bösen Tieren eingeholt und zu Boden geworfen werden – also beginne ich meinen Rückzug mit kleinen Schritten, den Vierbeiner immer im Blick, um im Notfall doch panikartig losrennen zu können. Als ich endlich auf dem Beifahrersitz außerhalb der Gefahr sitze, komme ich mir total bescheuert vor.

Ich sehe meine Kamera – das Licht, die Constellation, die Wüste – mein Fotografenherz brennt. Das Licht schwindet und die Anmut der Constellation wechselt fast im Sekundentakt. Mein Vierbeiner ist jetzt nahe an meiner Kamera, schnüffelt und pinkelt in den Wüstensand. Die Stimmung bleibt friedlich – so auch das Tier.

Also steige ich wieder aus und gehe zu meiner Kamera, um weiterzuarbeiten. Ich versuche, die Gefahr zu ignorieren und stelle fest, dass die Gefahr nicht existiert. Mein Vierbeiner kommt nie näher als 10 bis 15 m an mich heran, schnüffelt, guckt mir ab und an beim Arbeiten und Belichten zu, strolcht herum, pinkelt wieder in den Sand – irgendwann sogar an „meine" Constellation. Wir sind sozusagen in einer friedlichen Koexistenz, die sogar in eine gegenseitige Sympathie umschlägt – für mich jedenfalls.

Ich fotografiere weiter die Szenerie, bis kein Licht mehr existiert und es kalt und dunkel ist. Mein Vierbeiner war irgendwann einfach weg – leider habe ich ihn nie fotografiert. Tage später erzählte ich meine Geschichte einem Mitarbeiter des Regionalflughafens. Es gäbe keine Huskys in der Wüste, meinte er, aber mit Sicherheit fotointeressierte Kojoten. Ich werde dieses Zusammentreffen zwischen Tier, Mensch und Maschine wohl nie vergessen.

Werner Bartsch

Cockpit

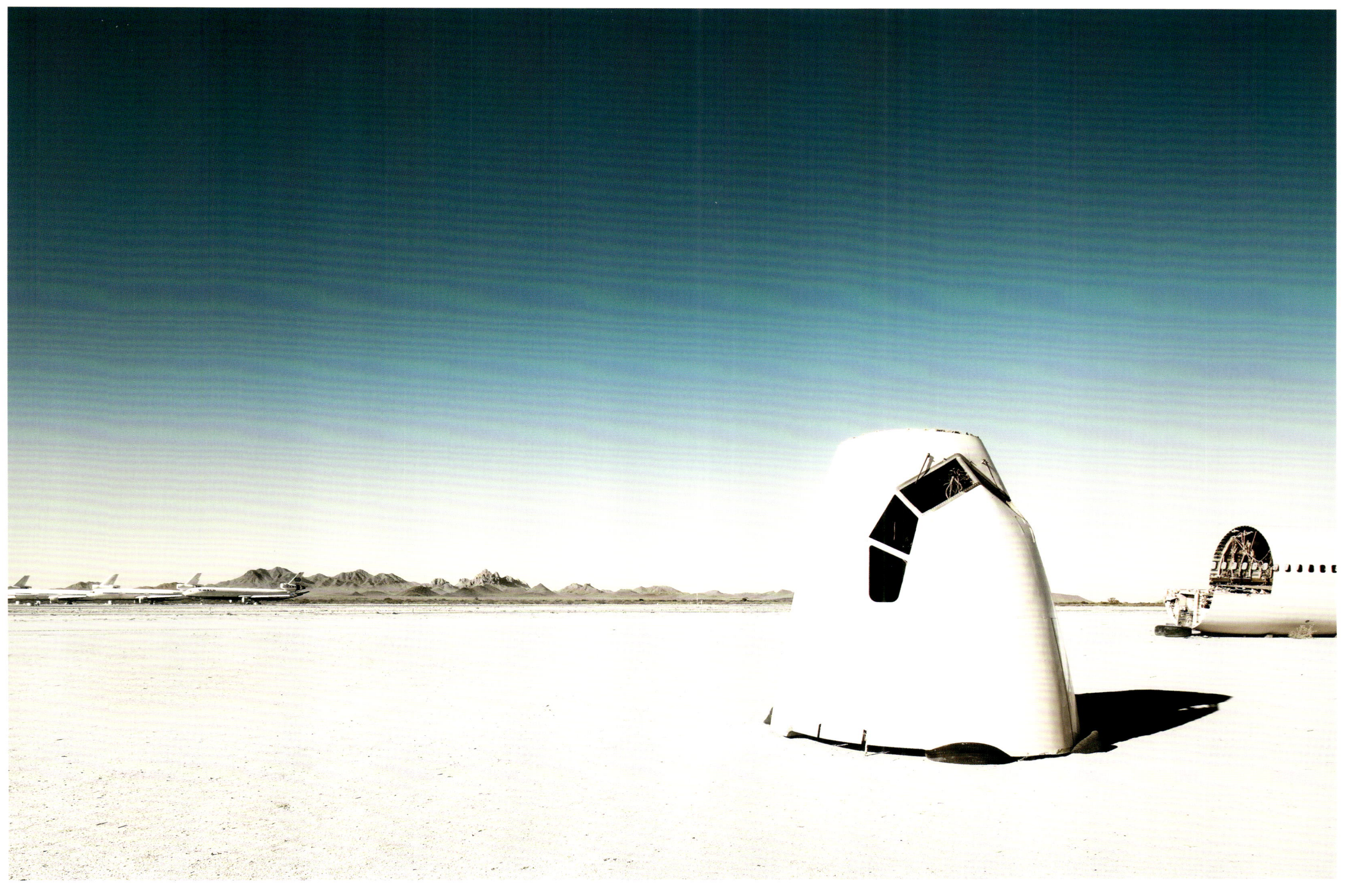

Tail #1

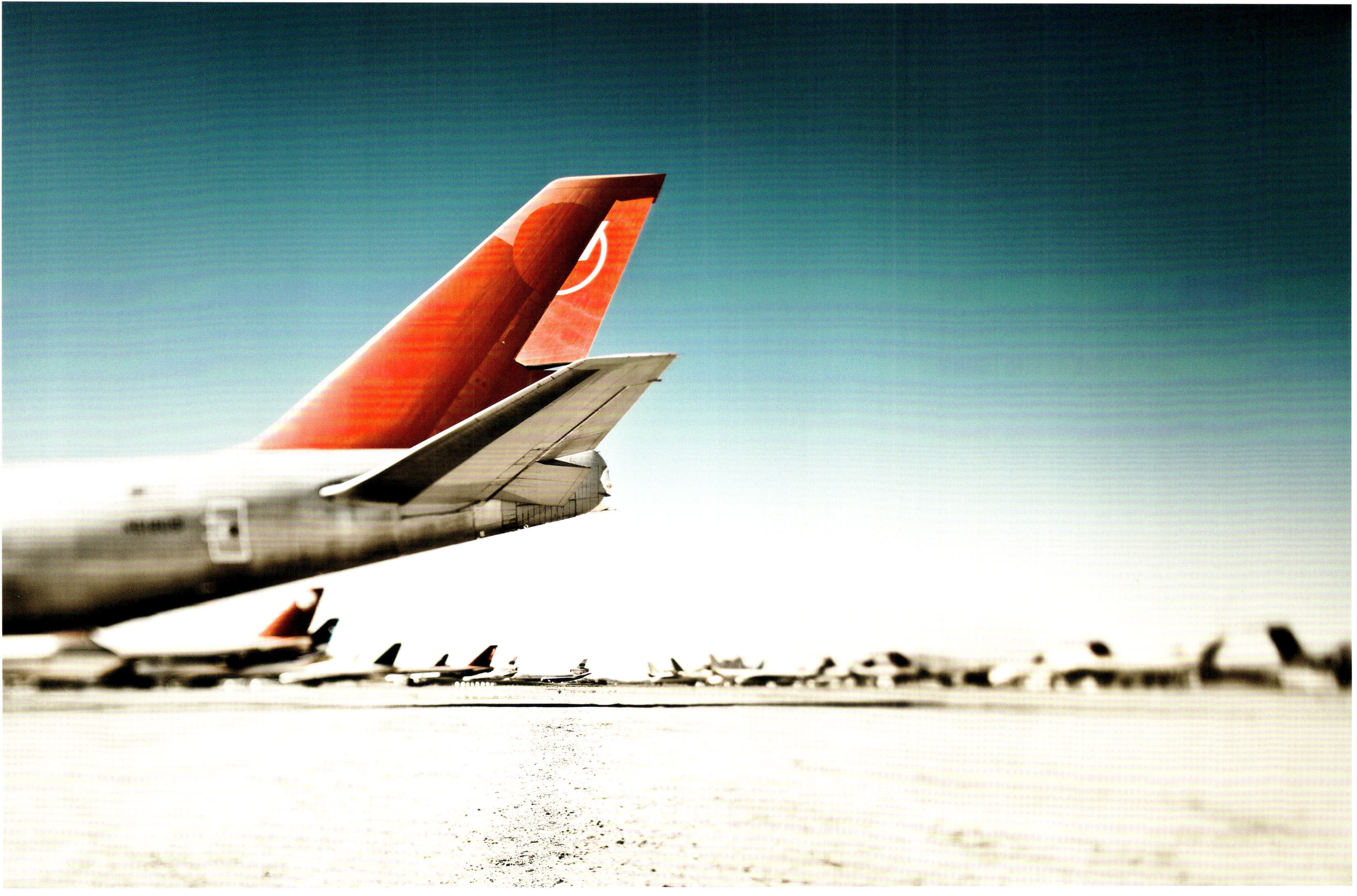

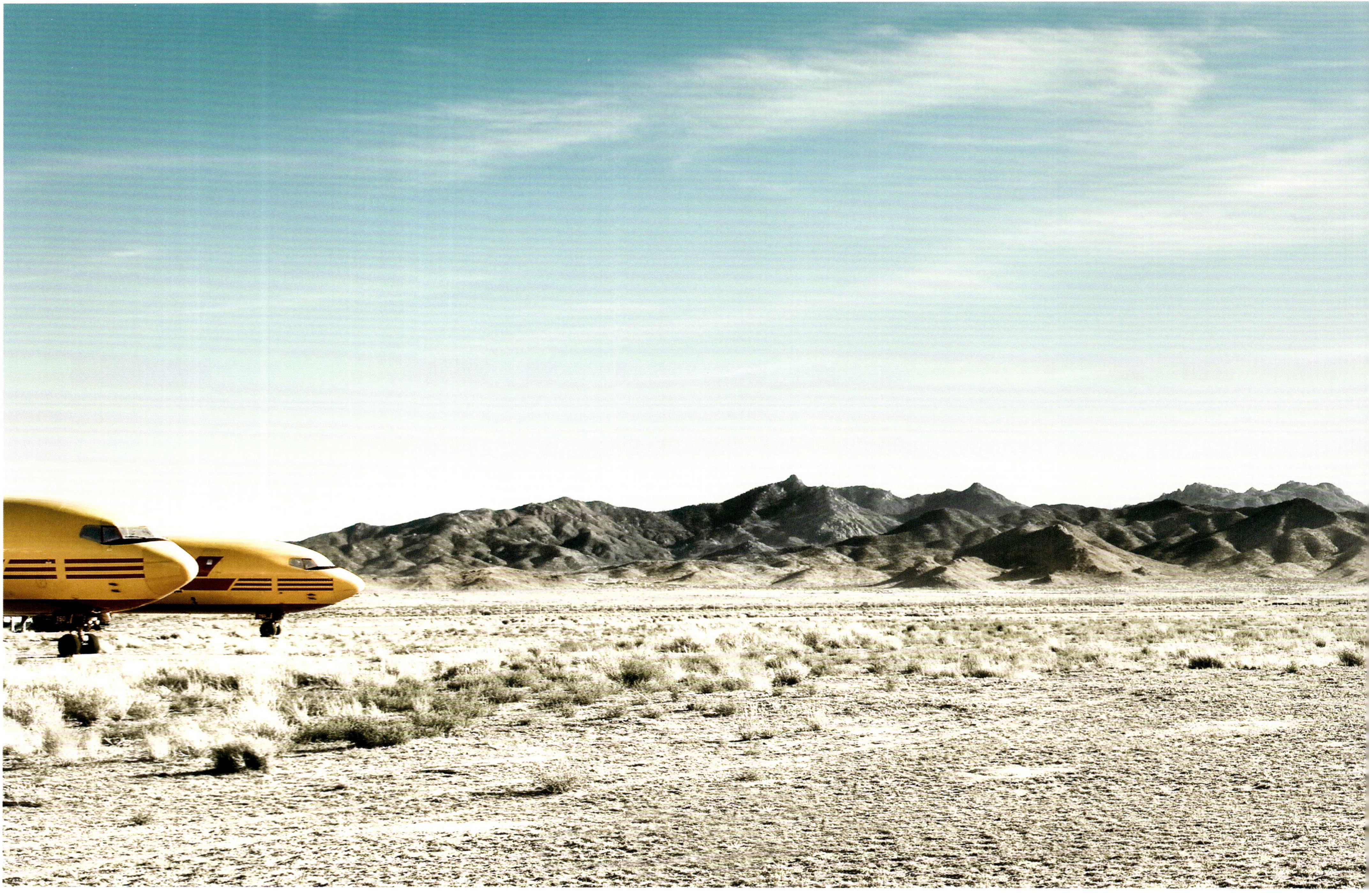

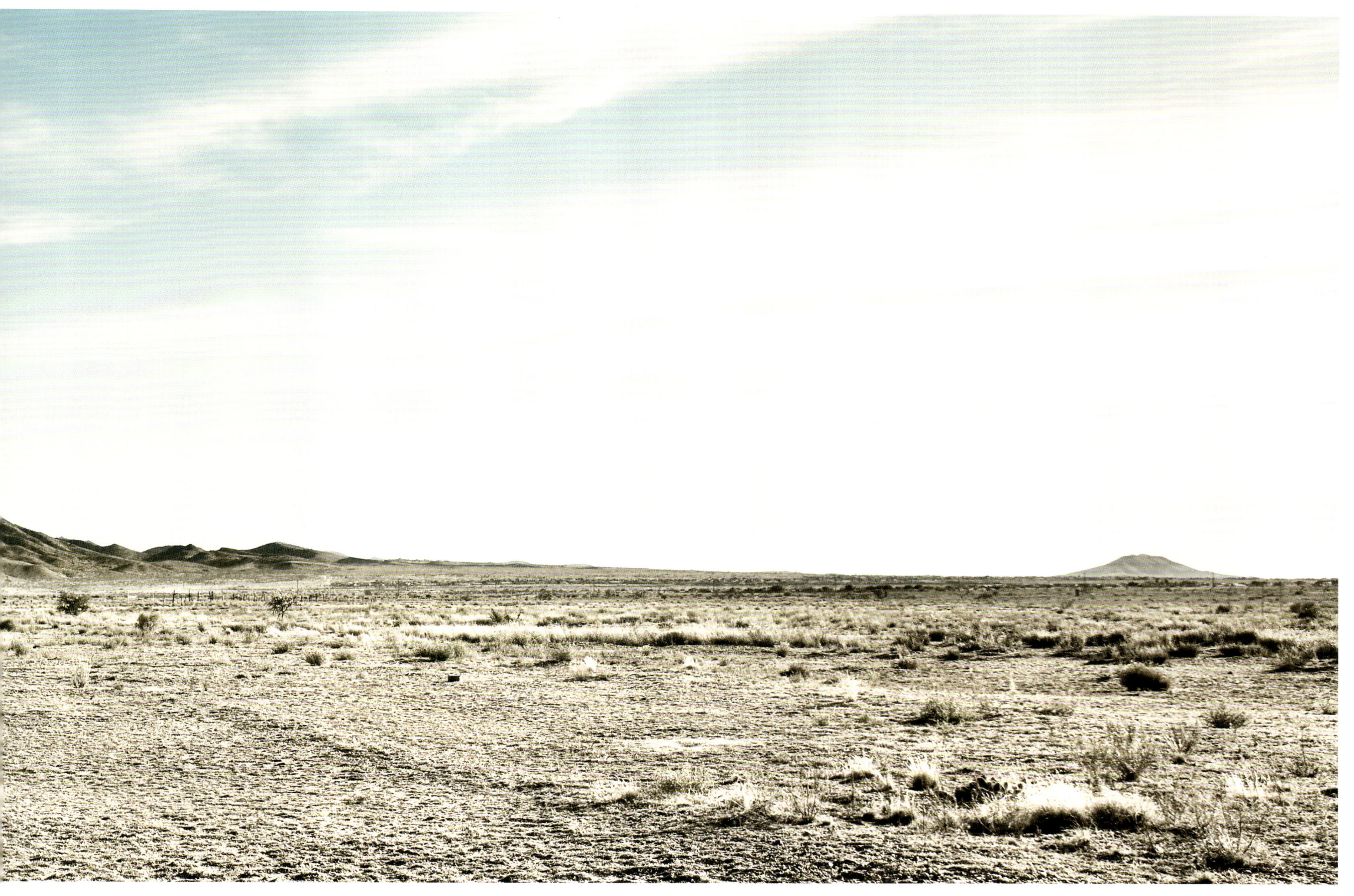

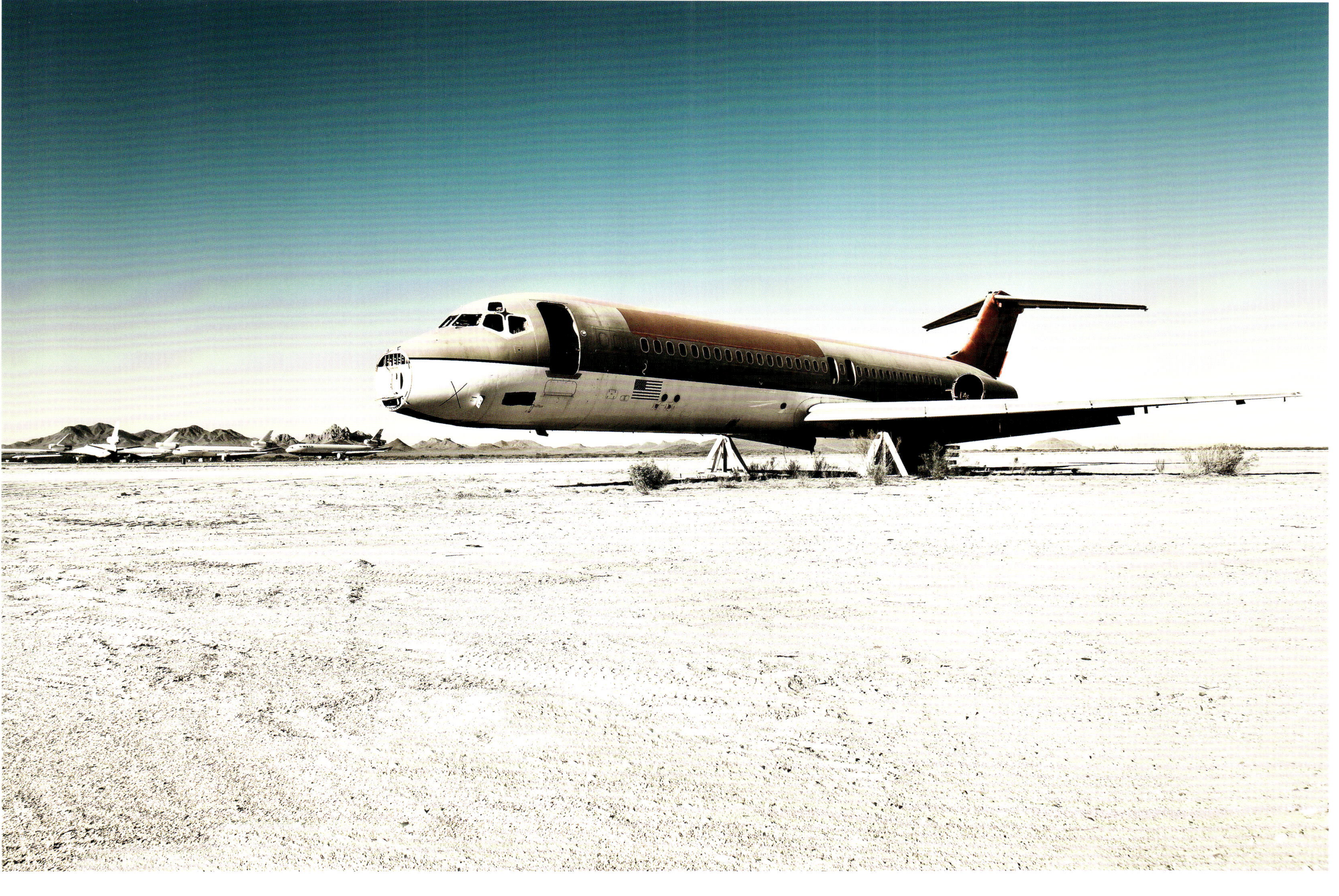

Gangway

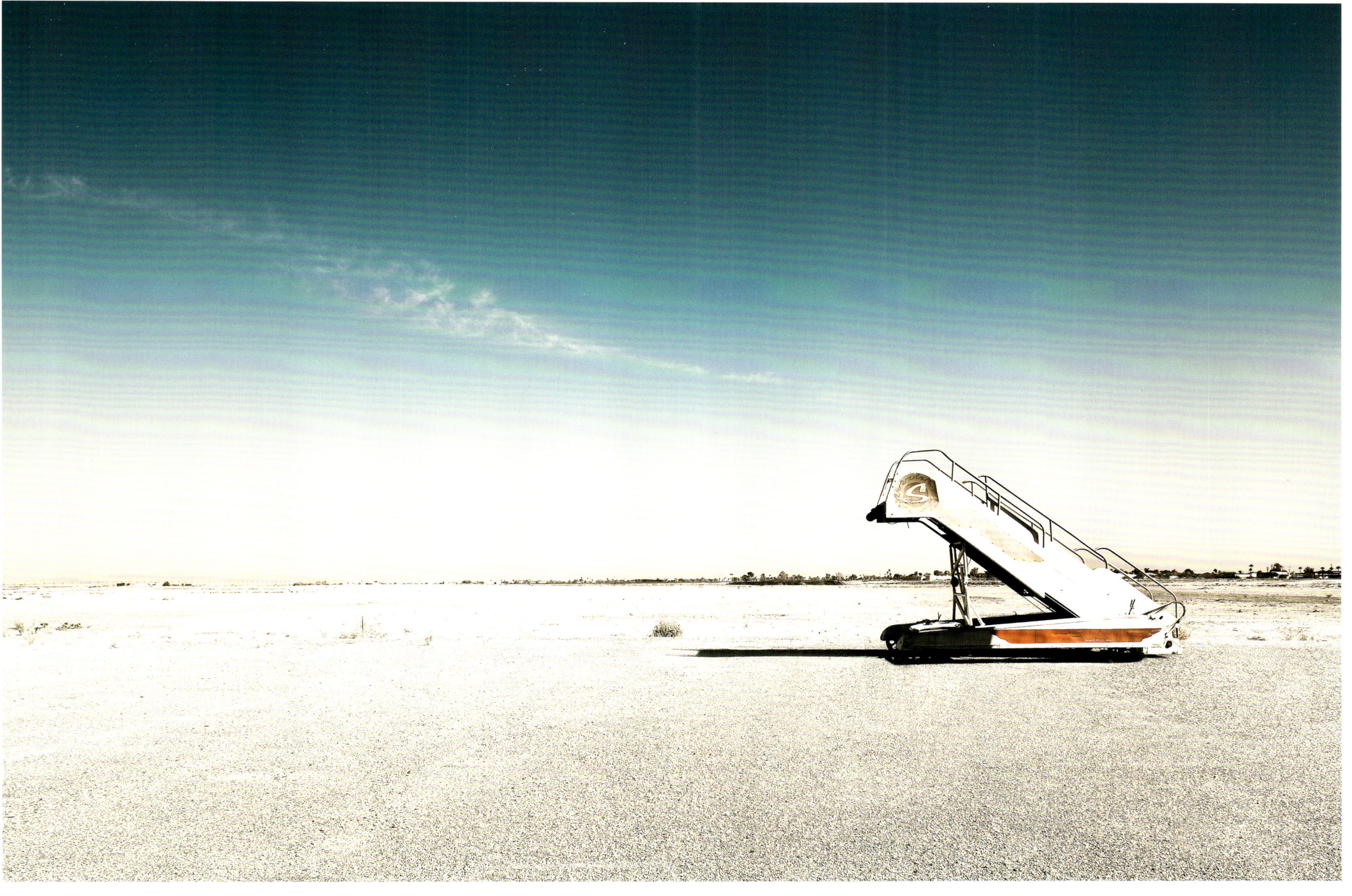

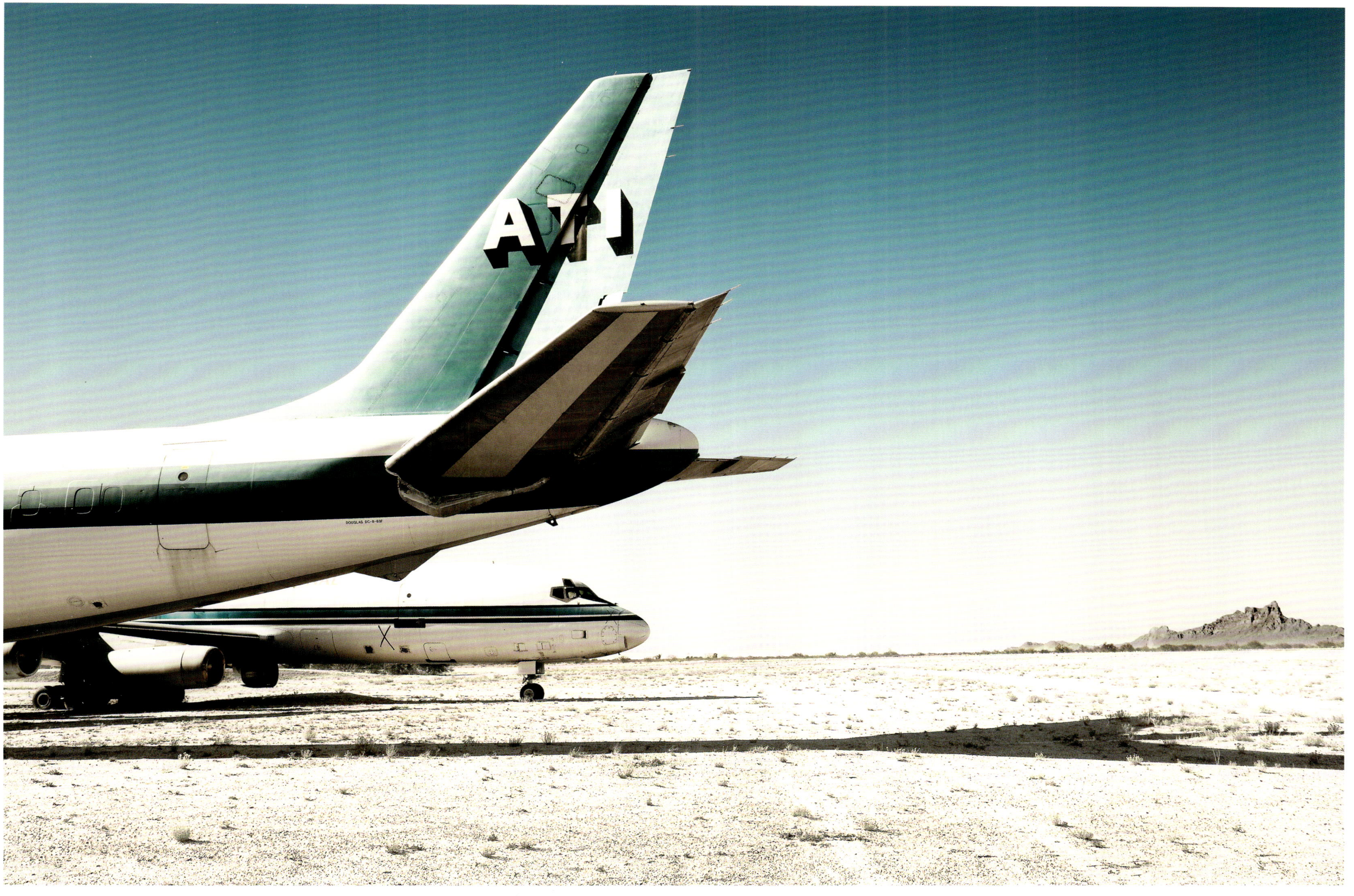
ATI
DOUGLAS DC-8-83F

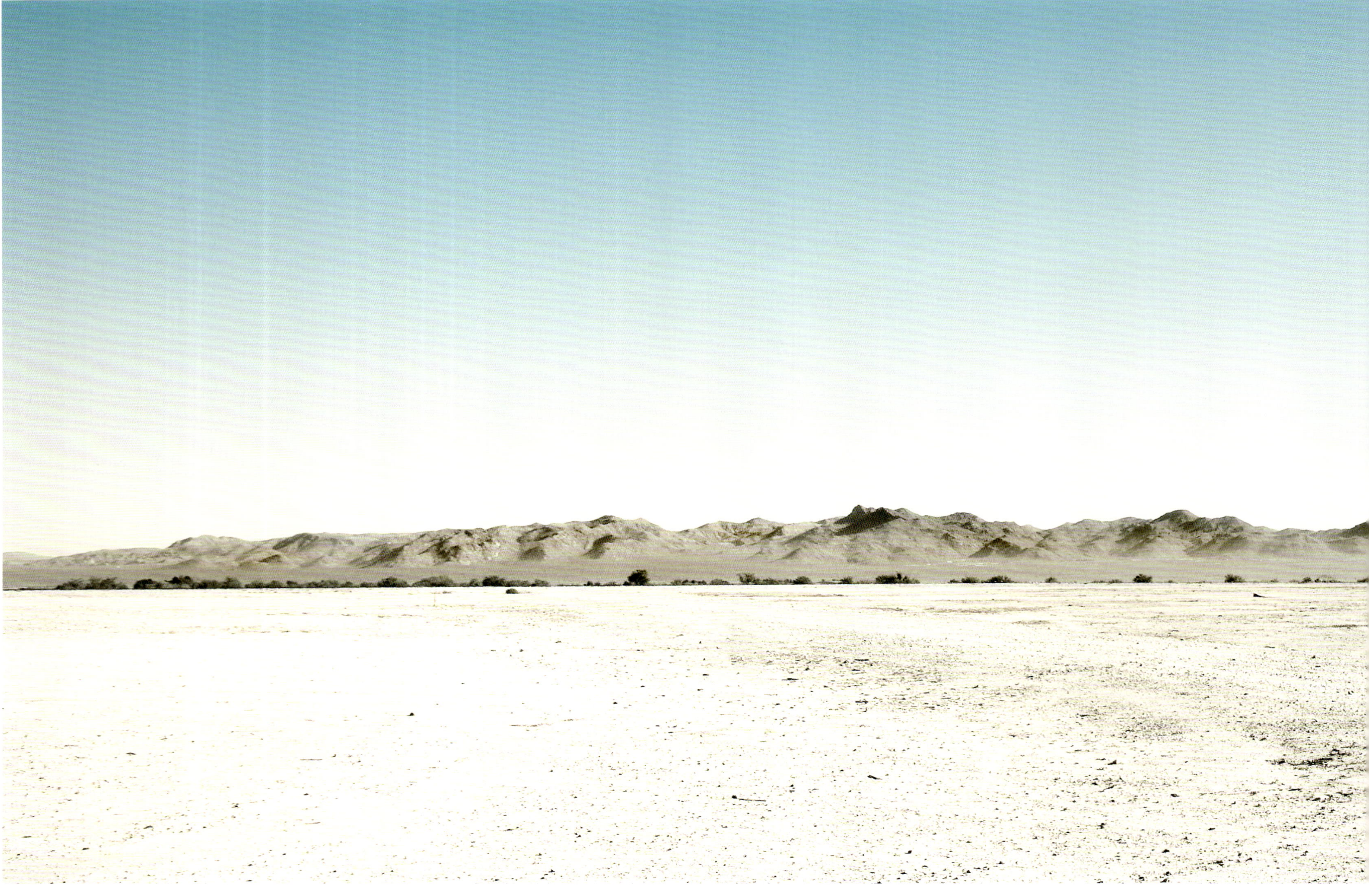

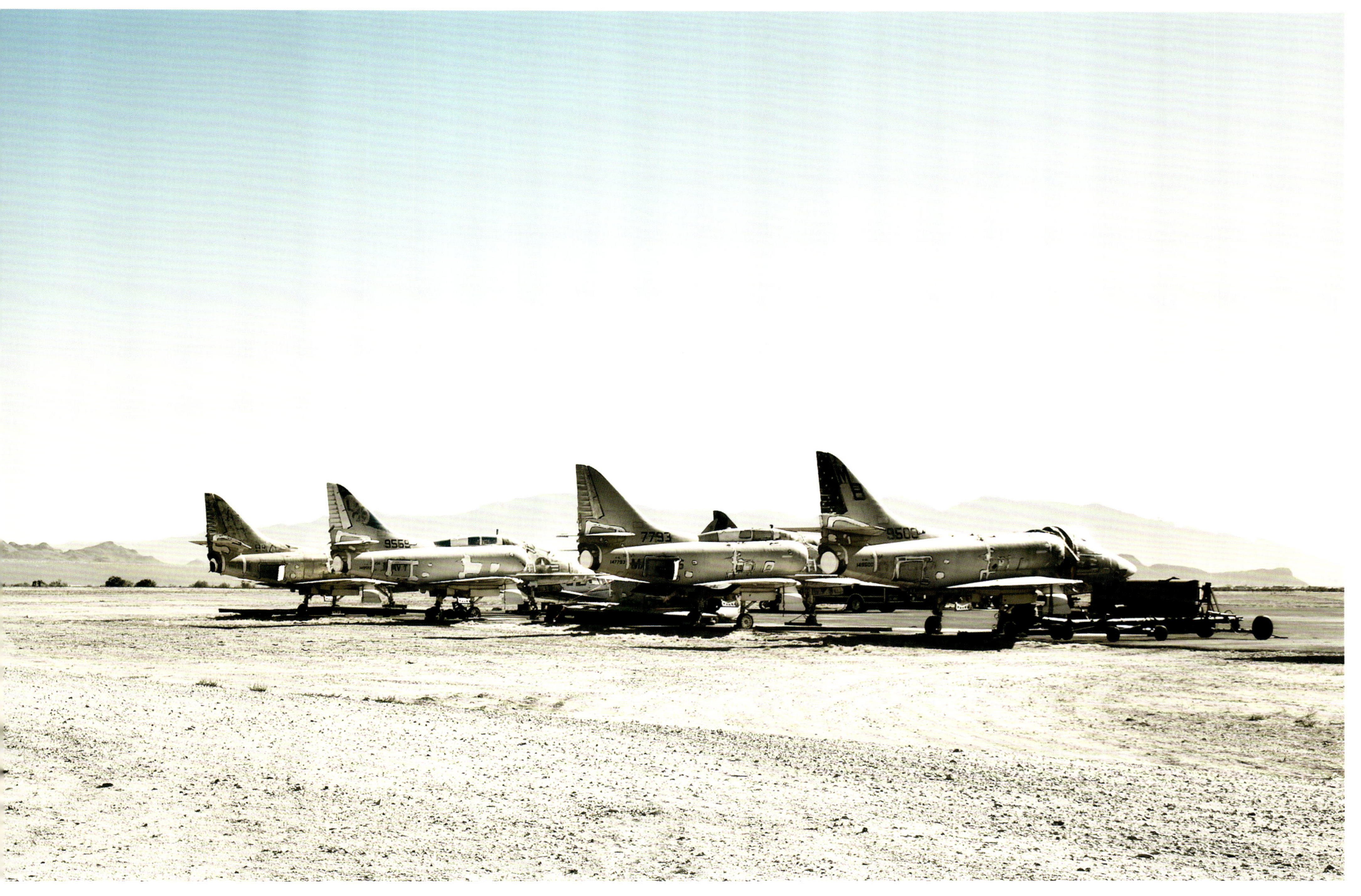

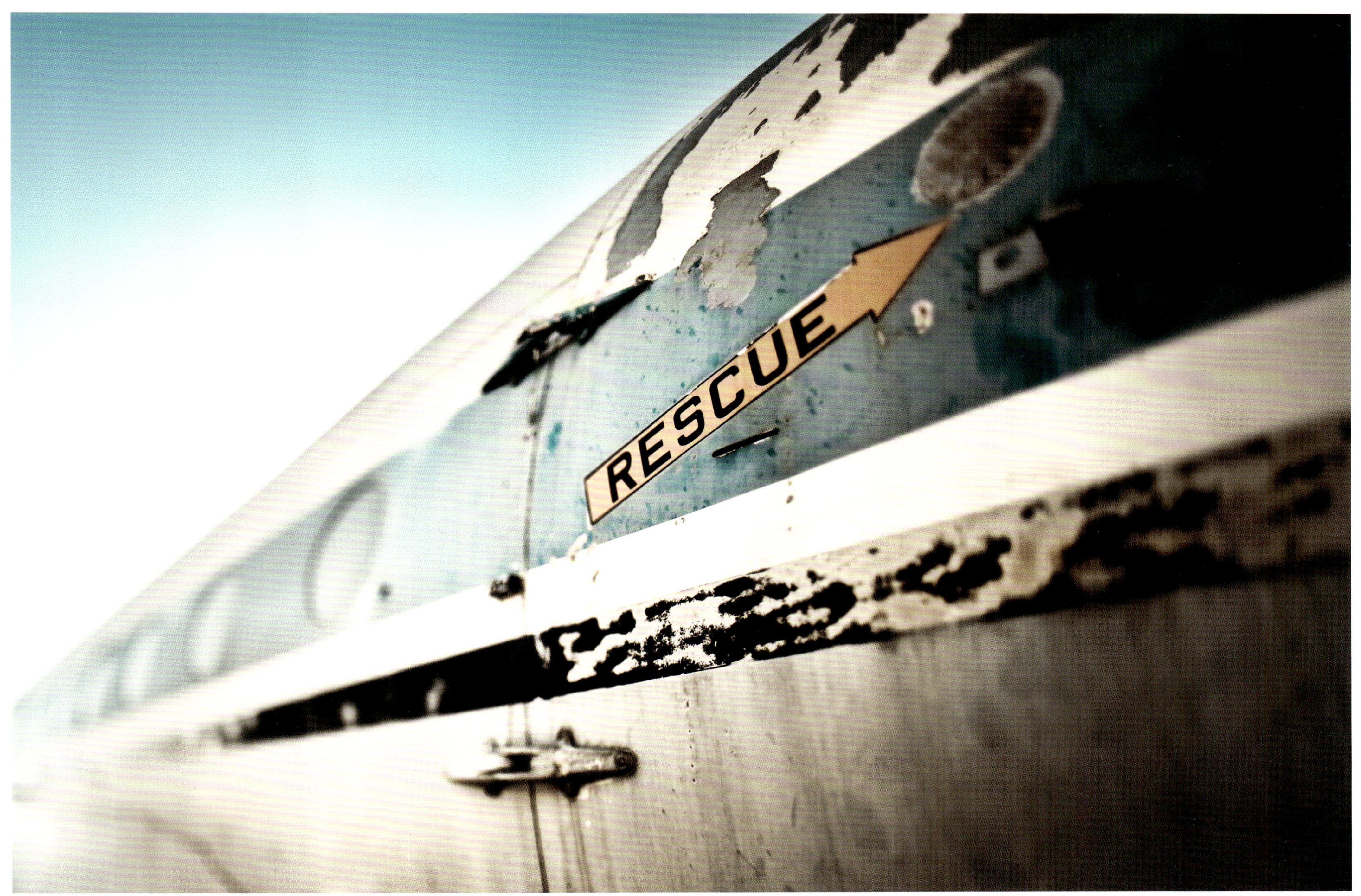
RESCUE

Taxiway

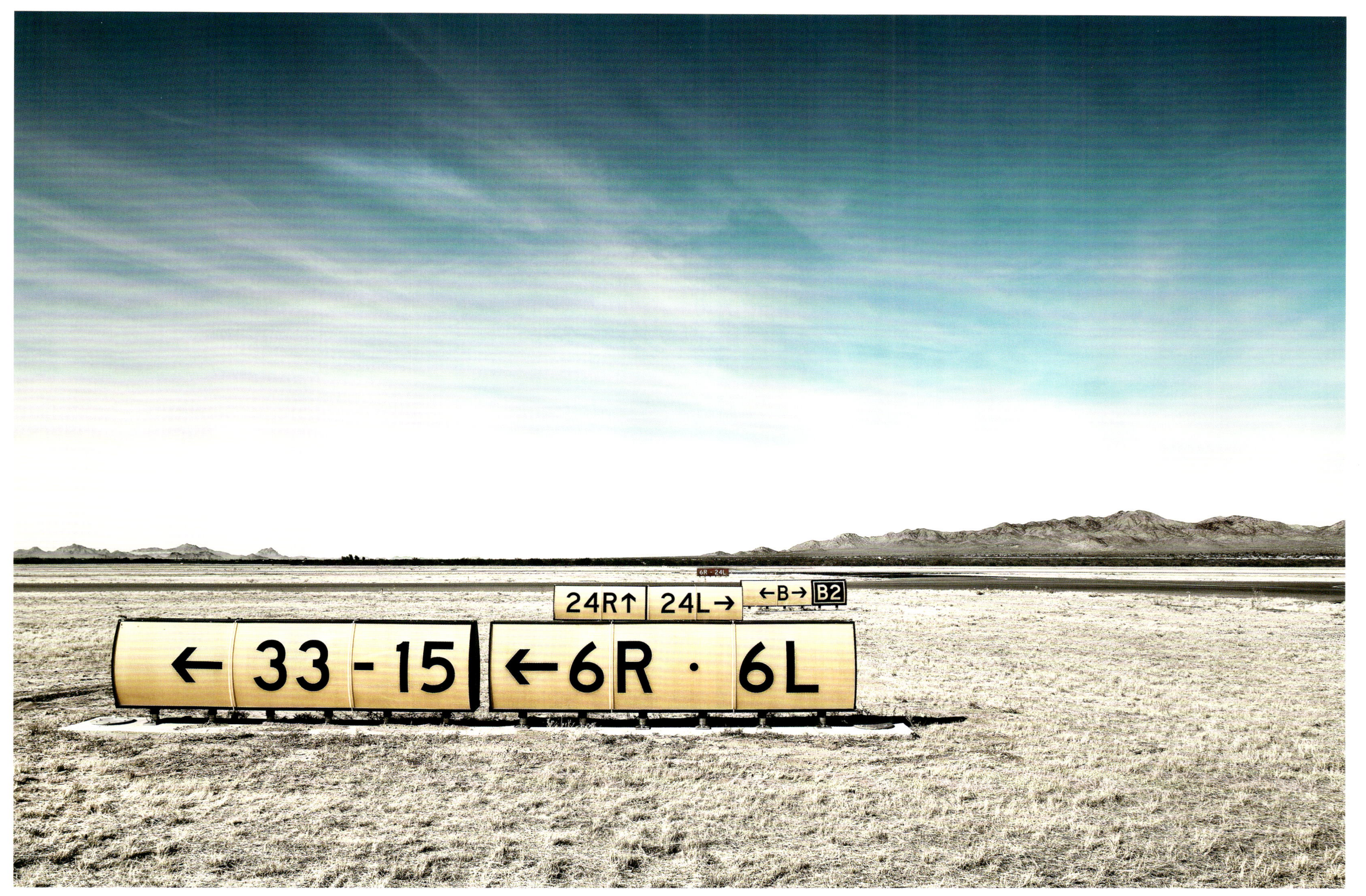
6R - 24L
24R↑ 24L→
←B→ B2
←33 - 15
←6R · 6L

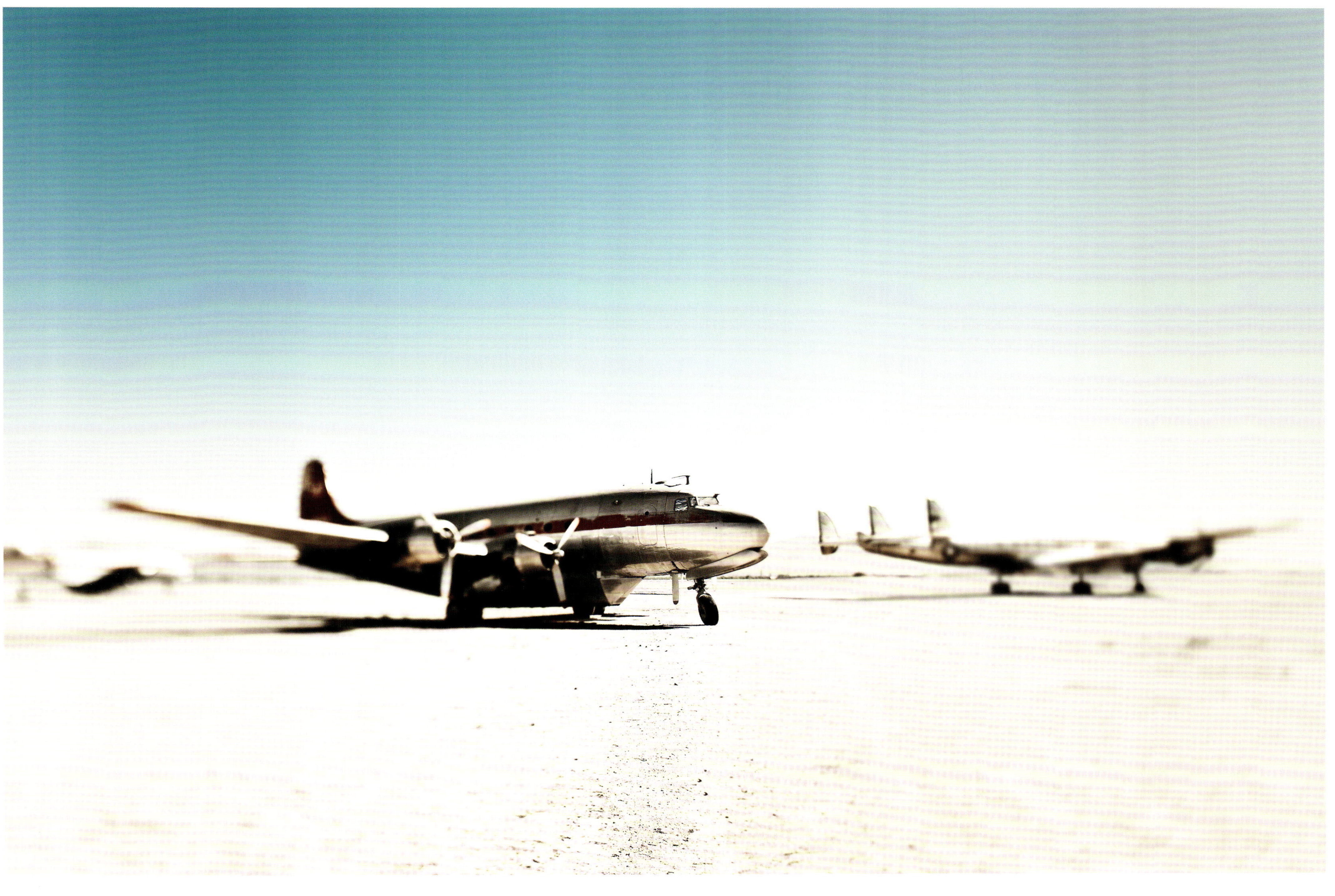

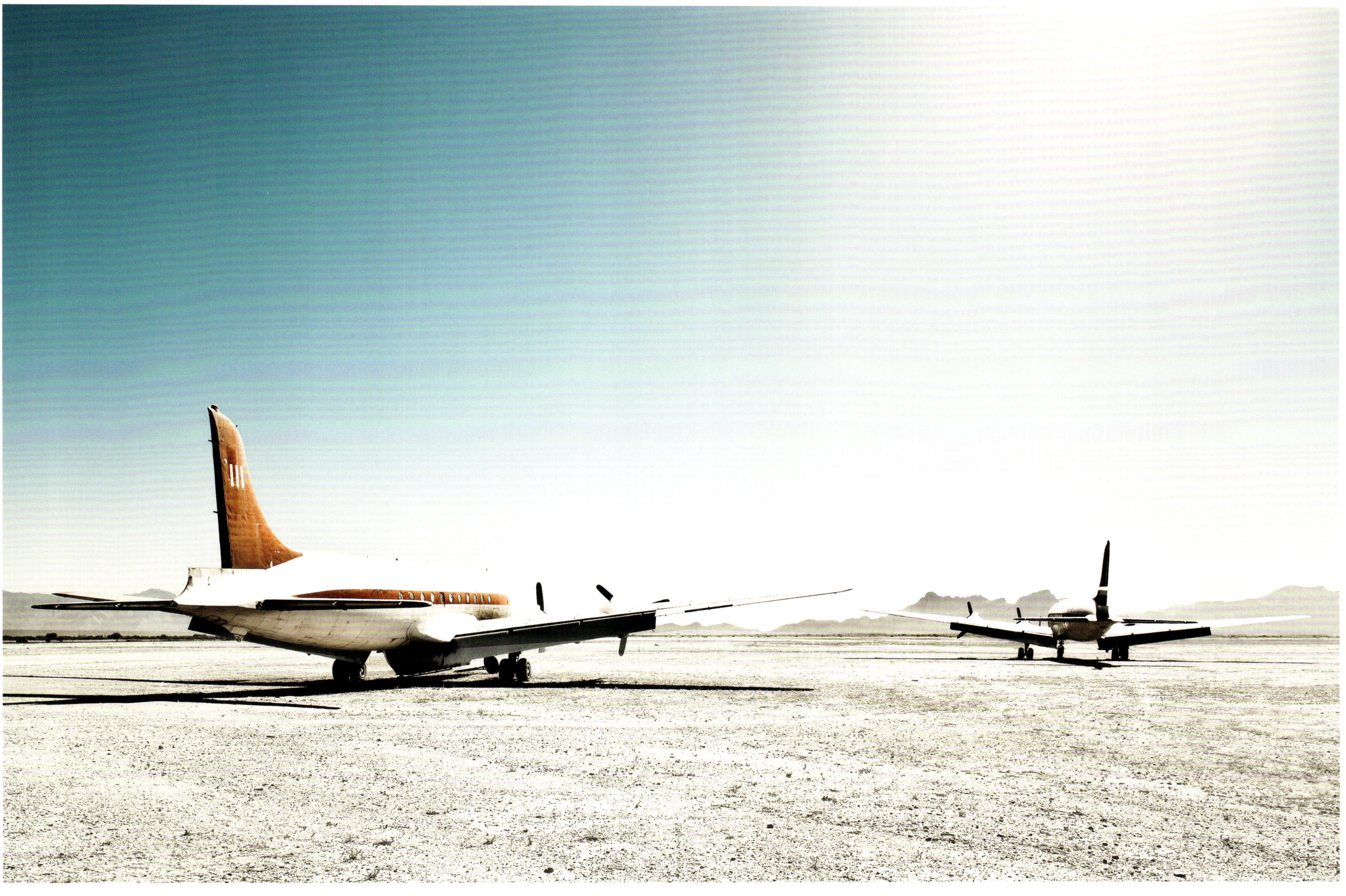

Full Moon #1

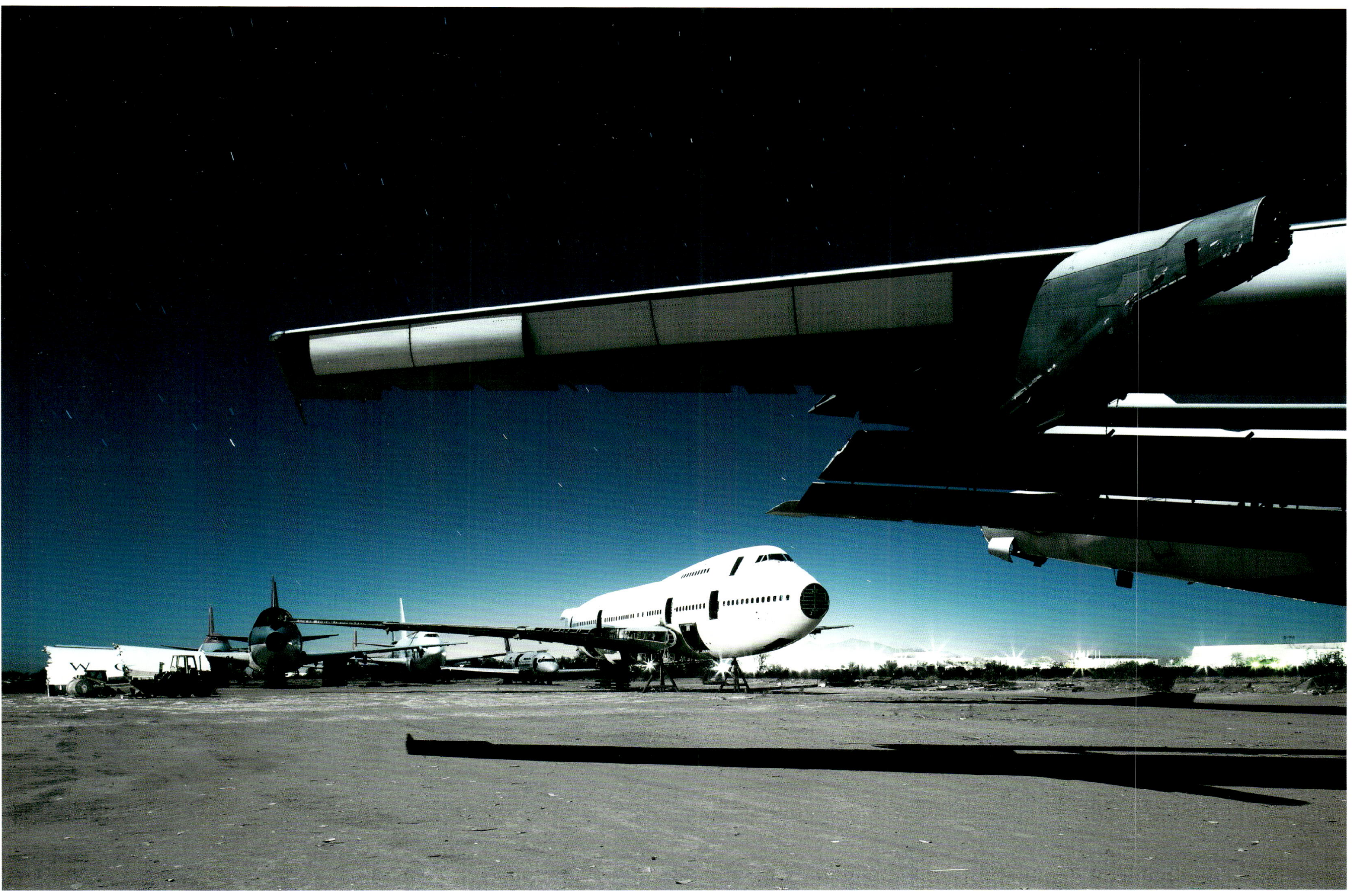

Full Moon #2

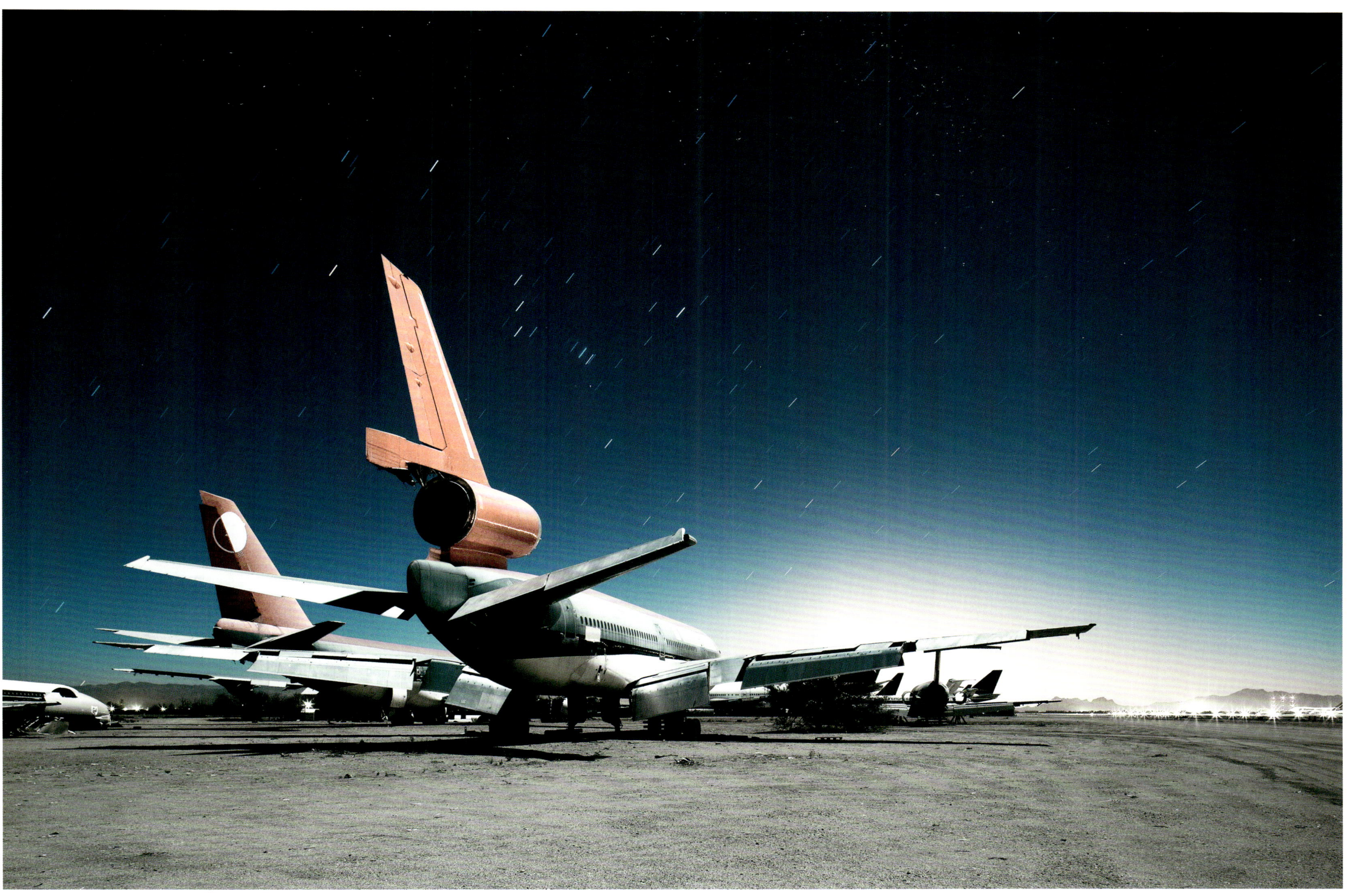

Full Moon #3

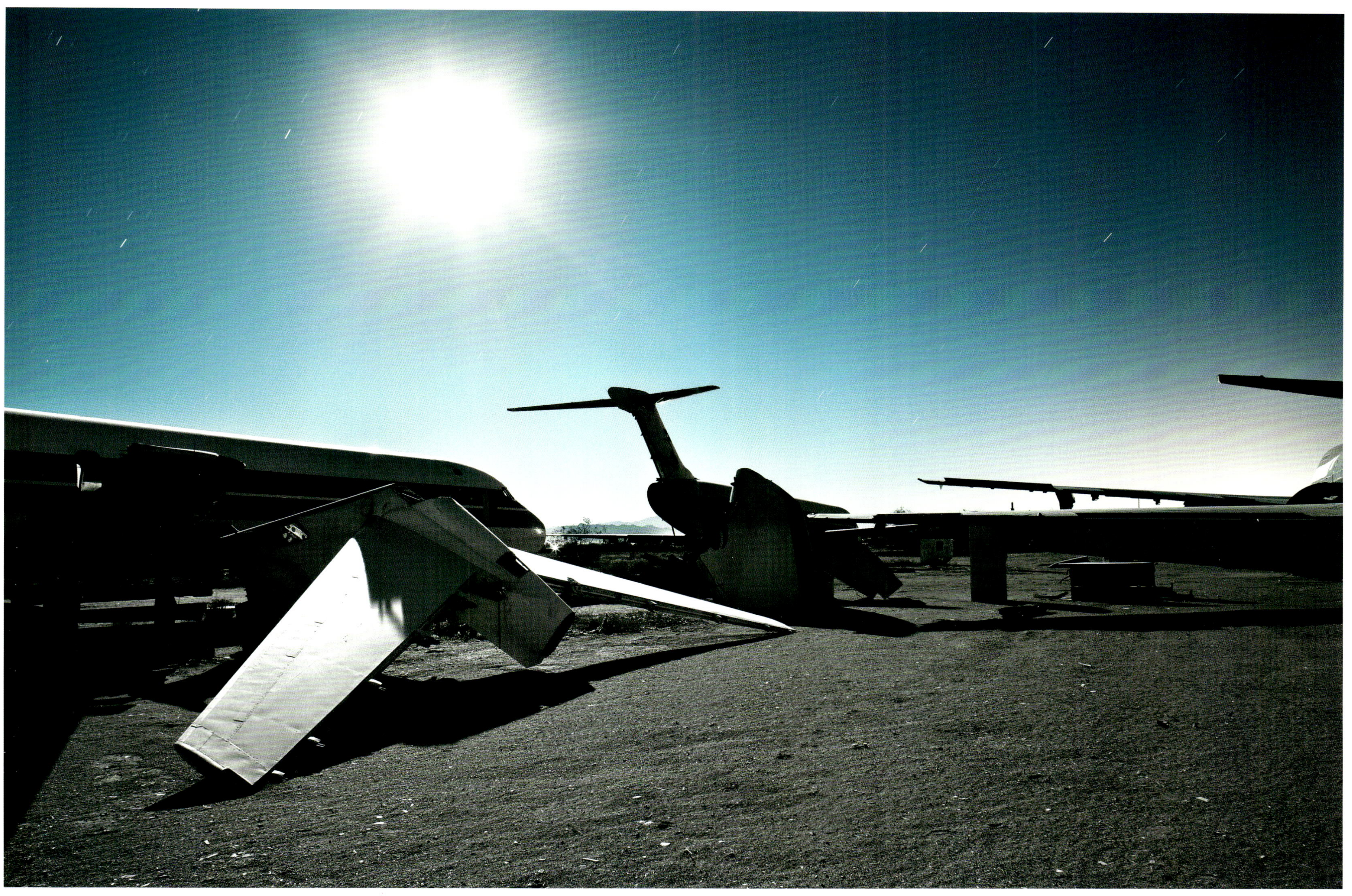

Full Moon #4

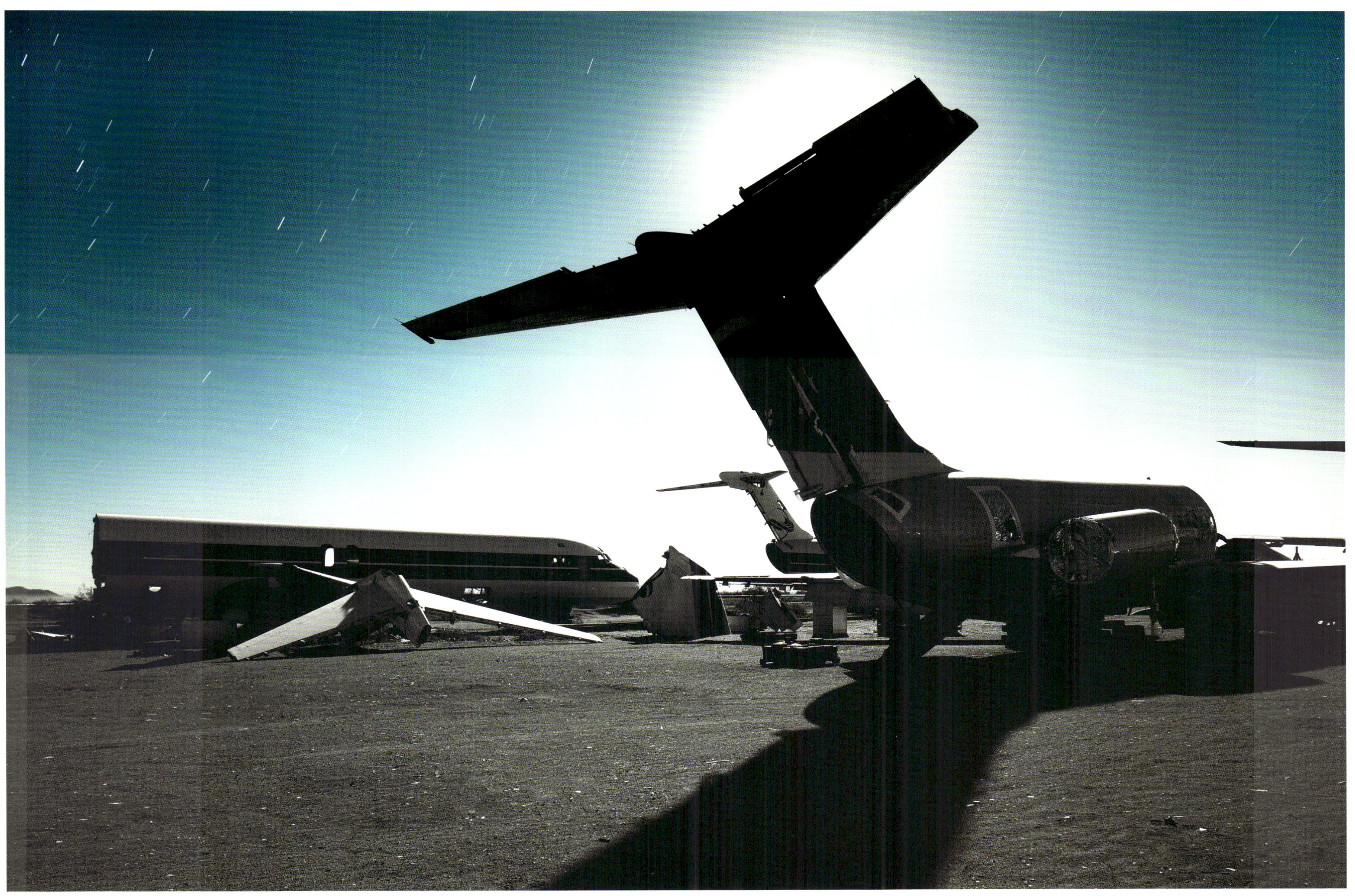

040

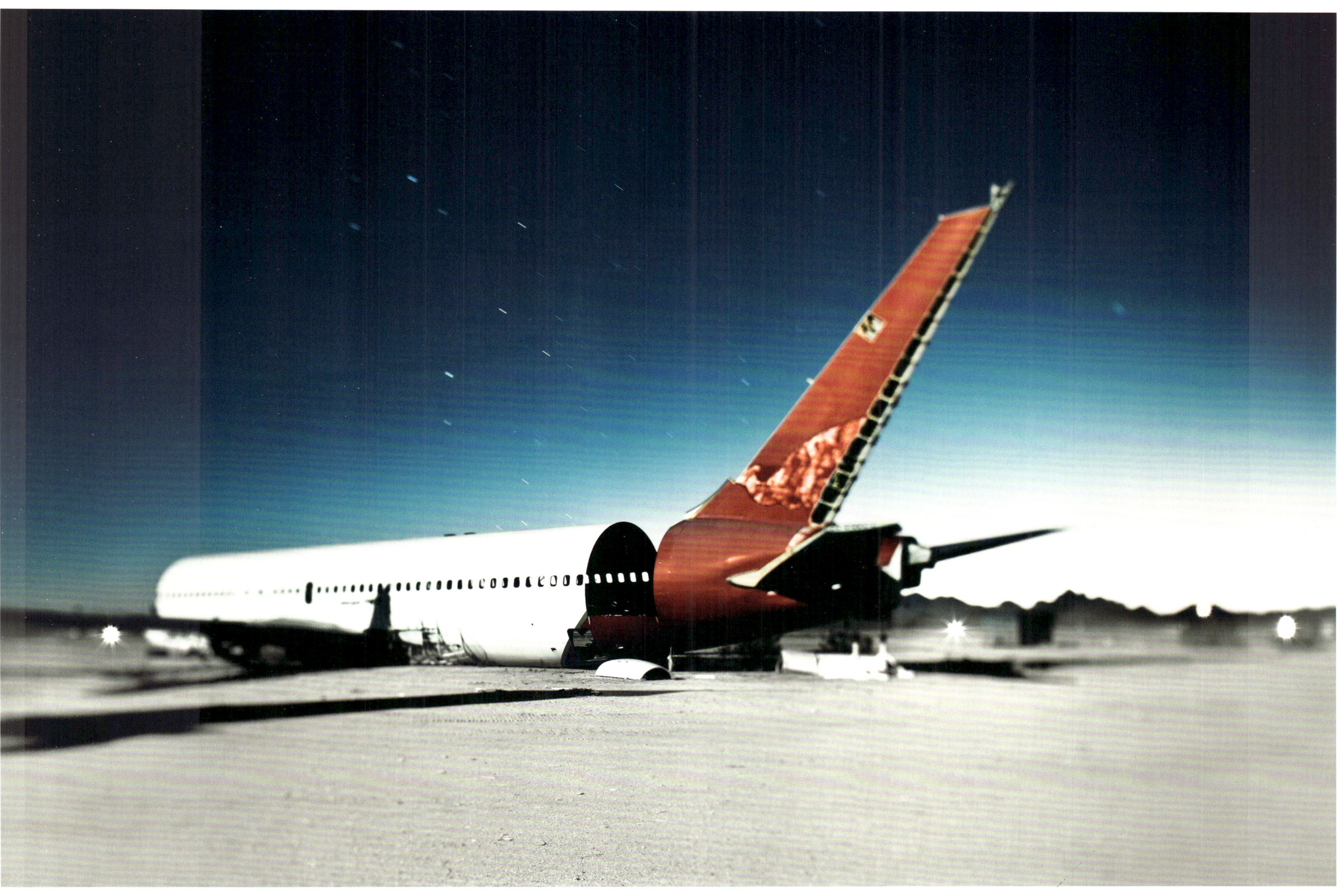

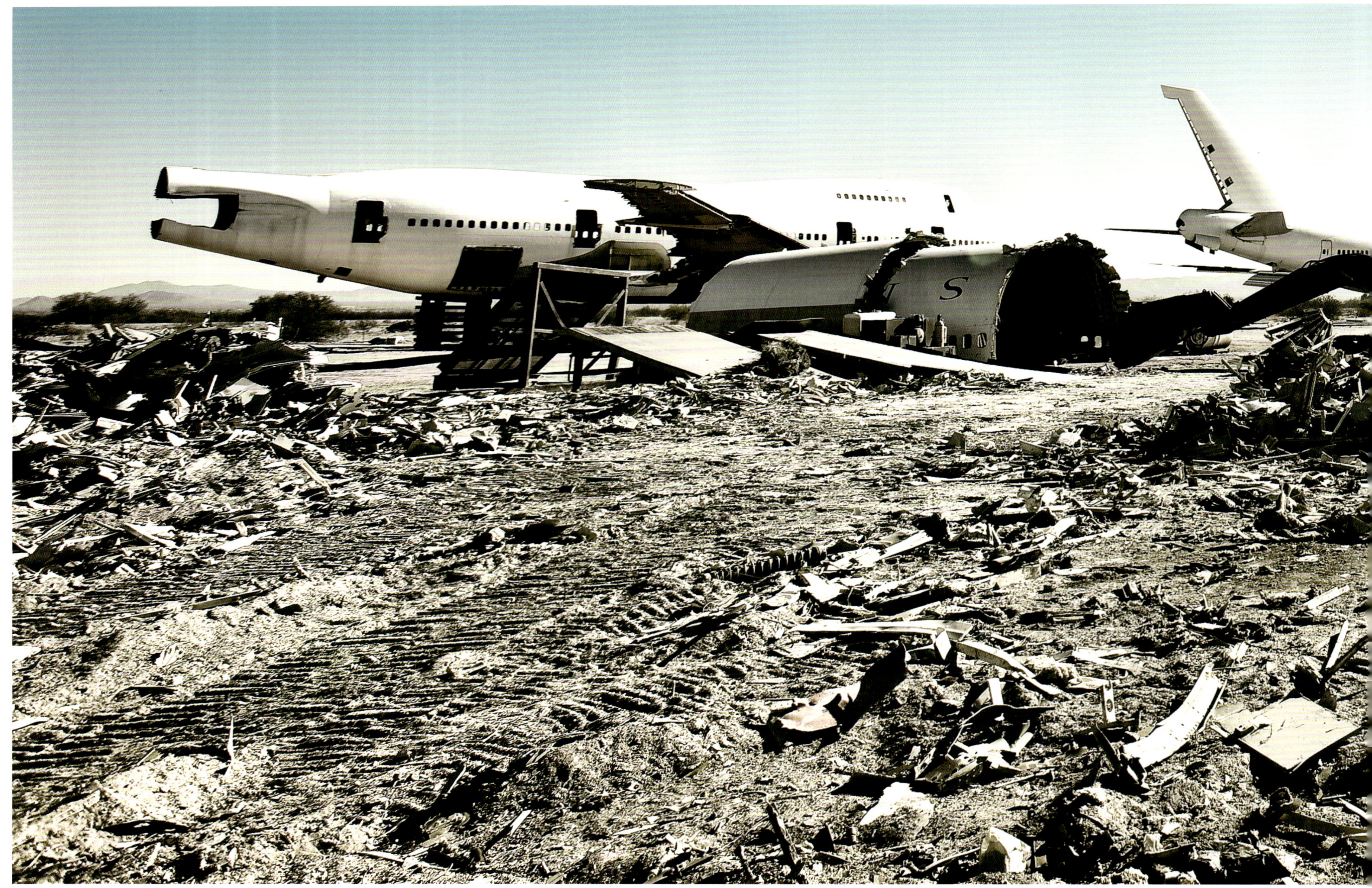

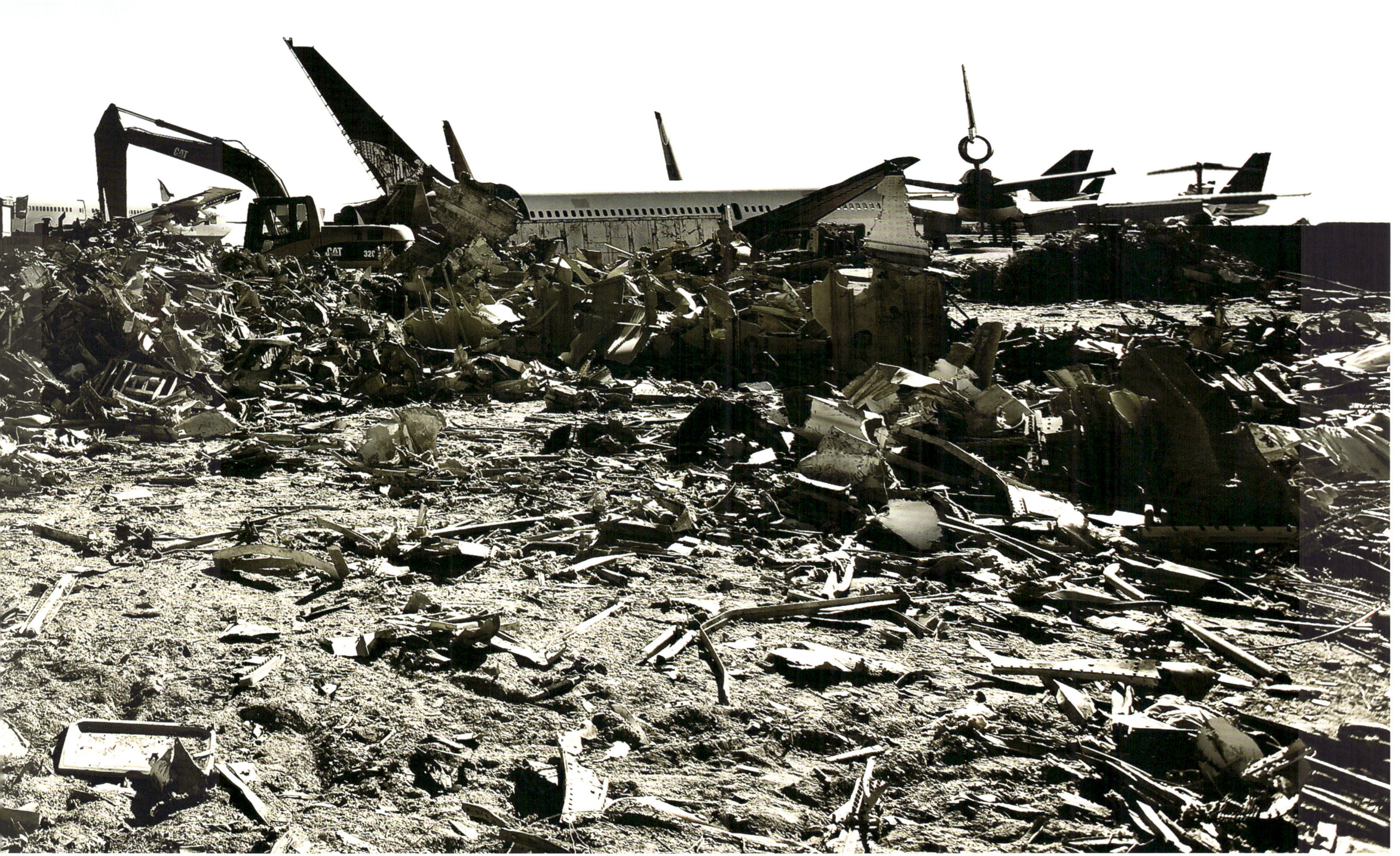

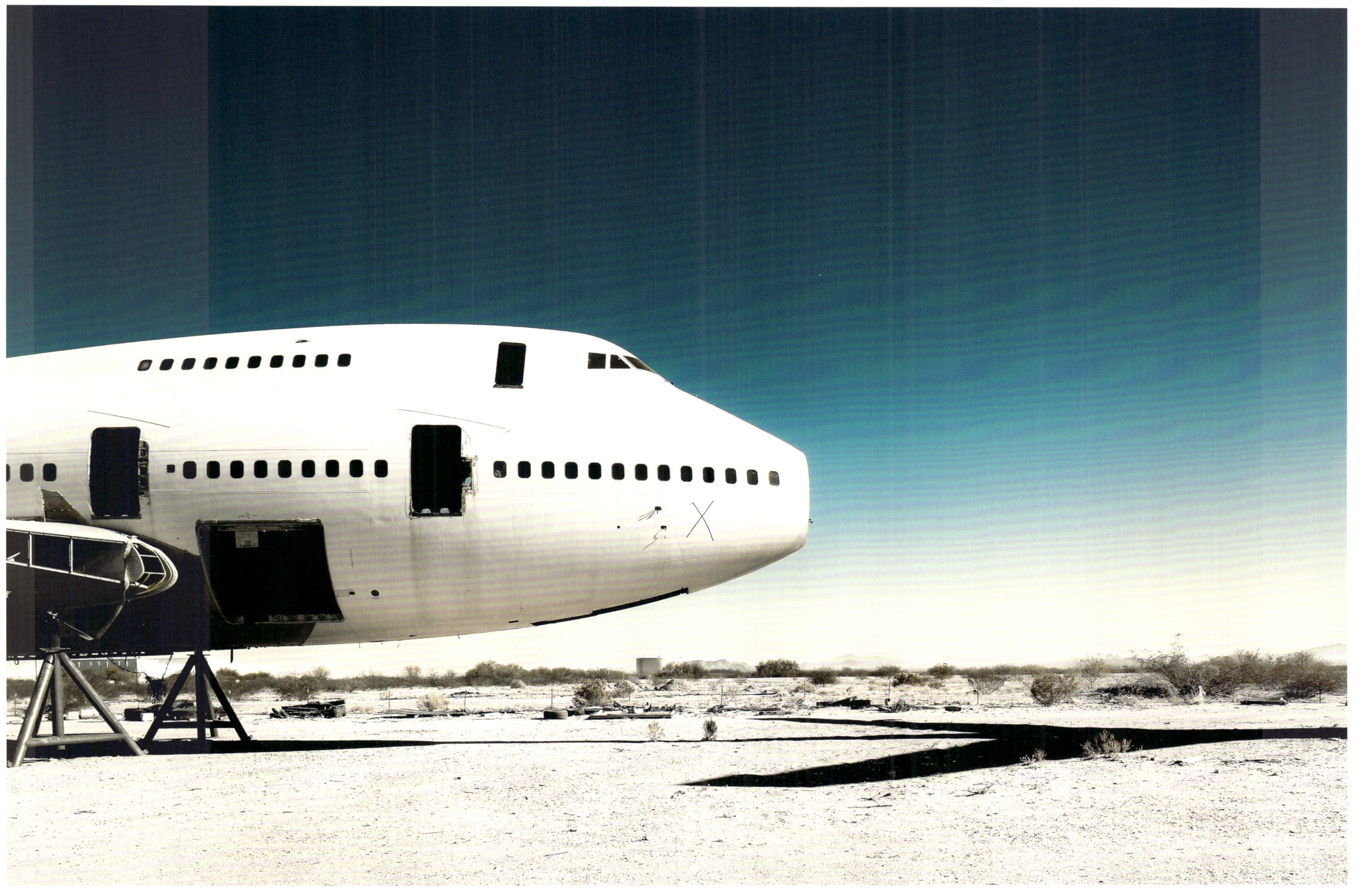

Boneyard #2

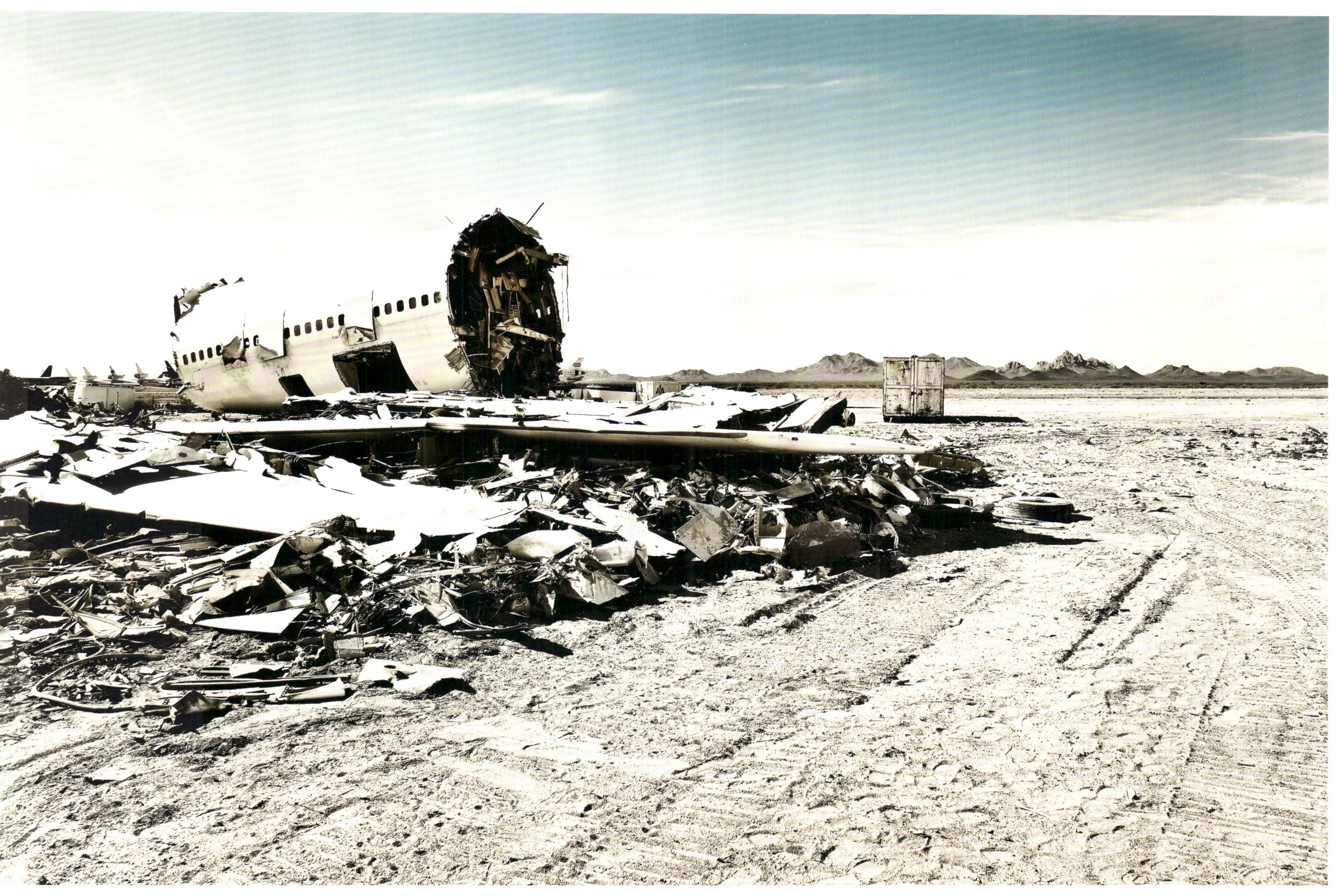

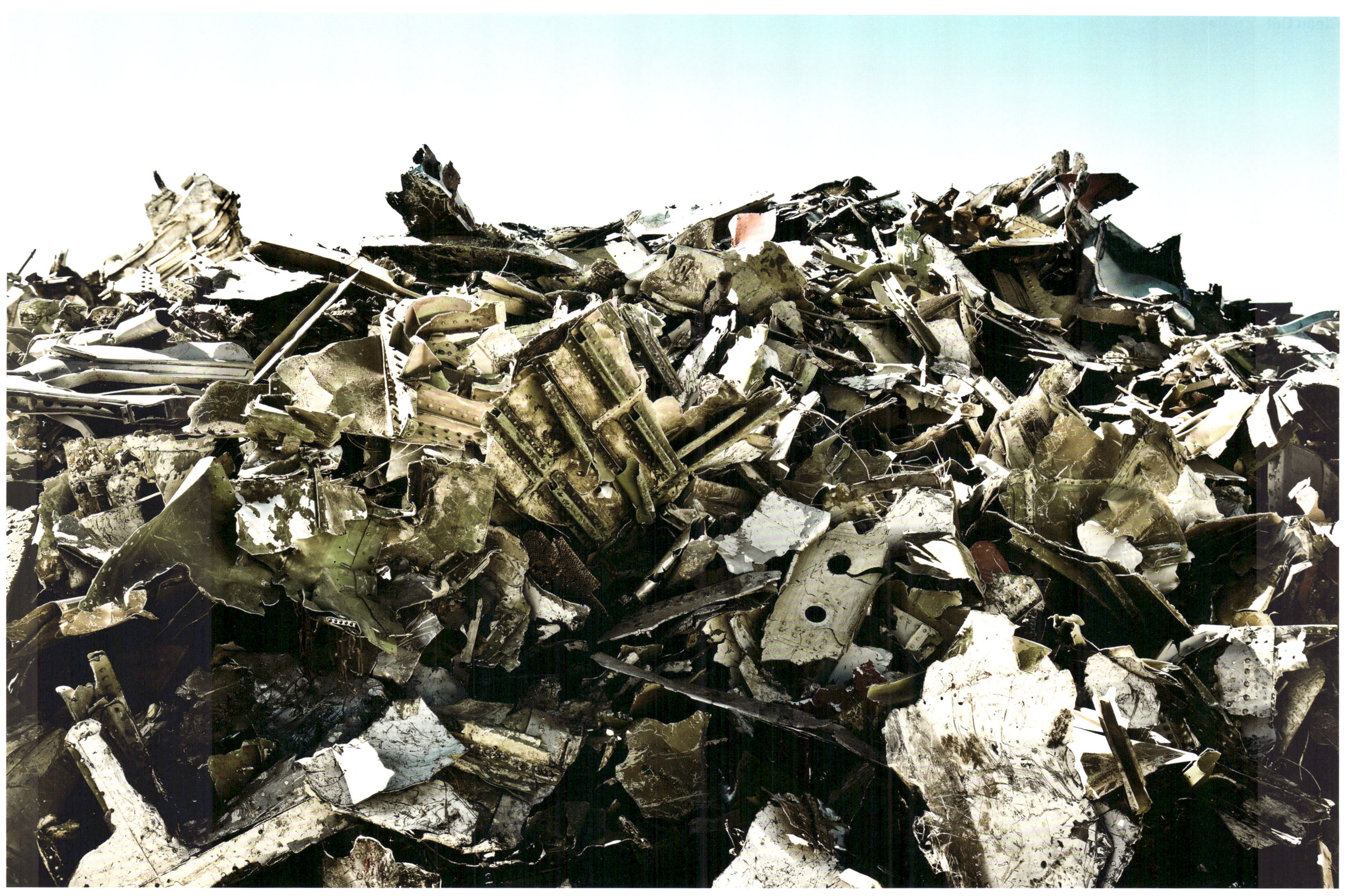

Clean Cut

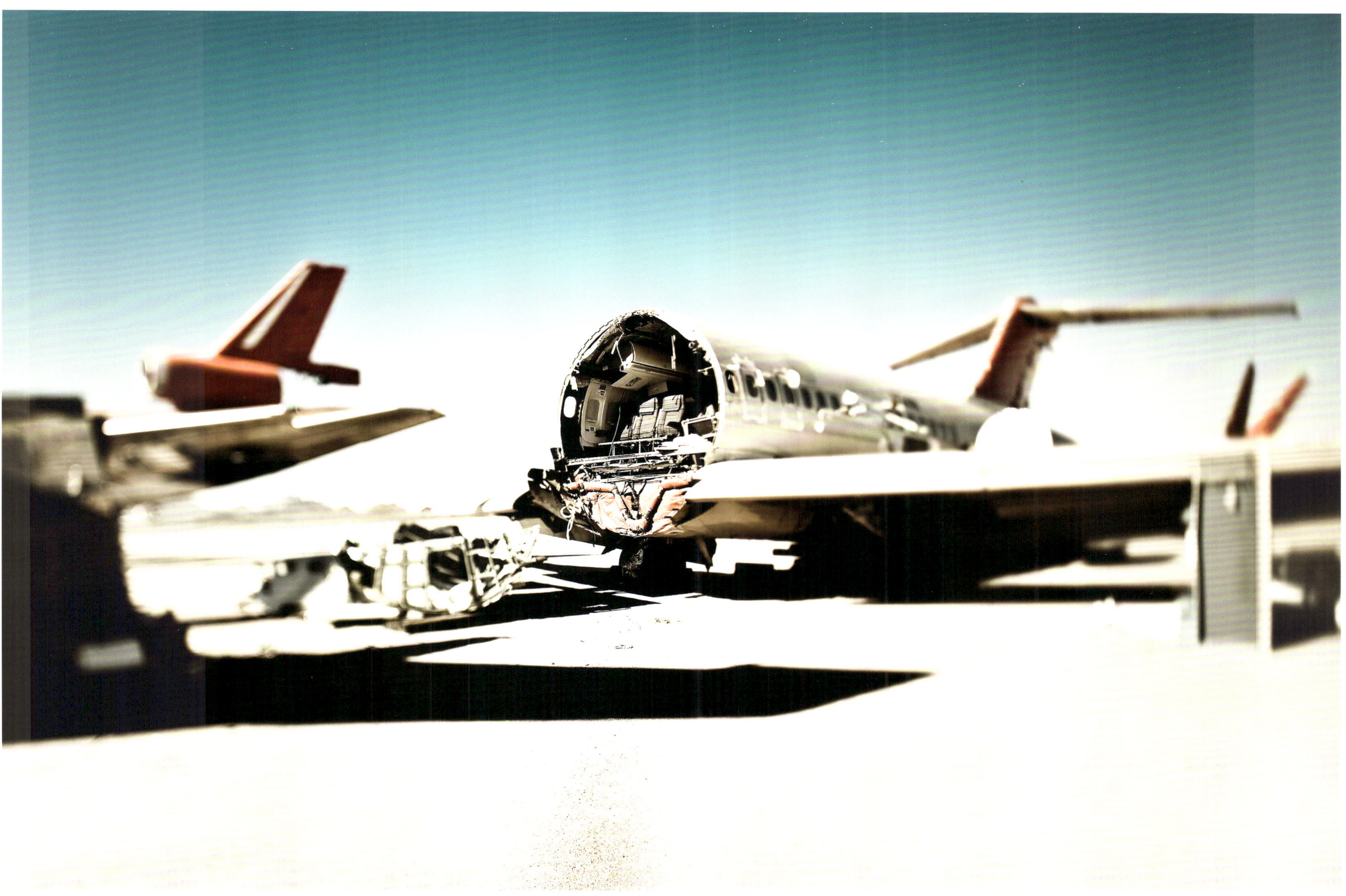

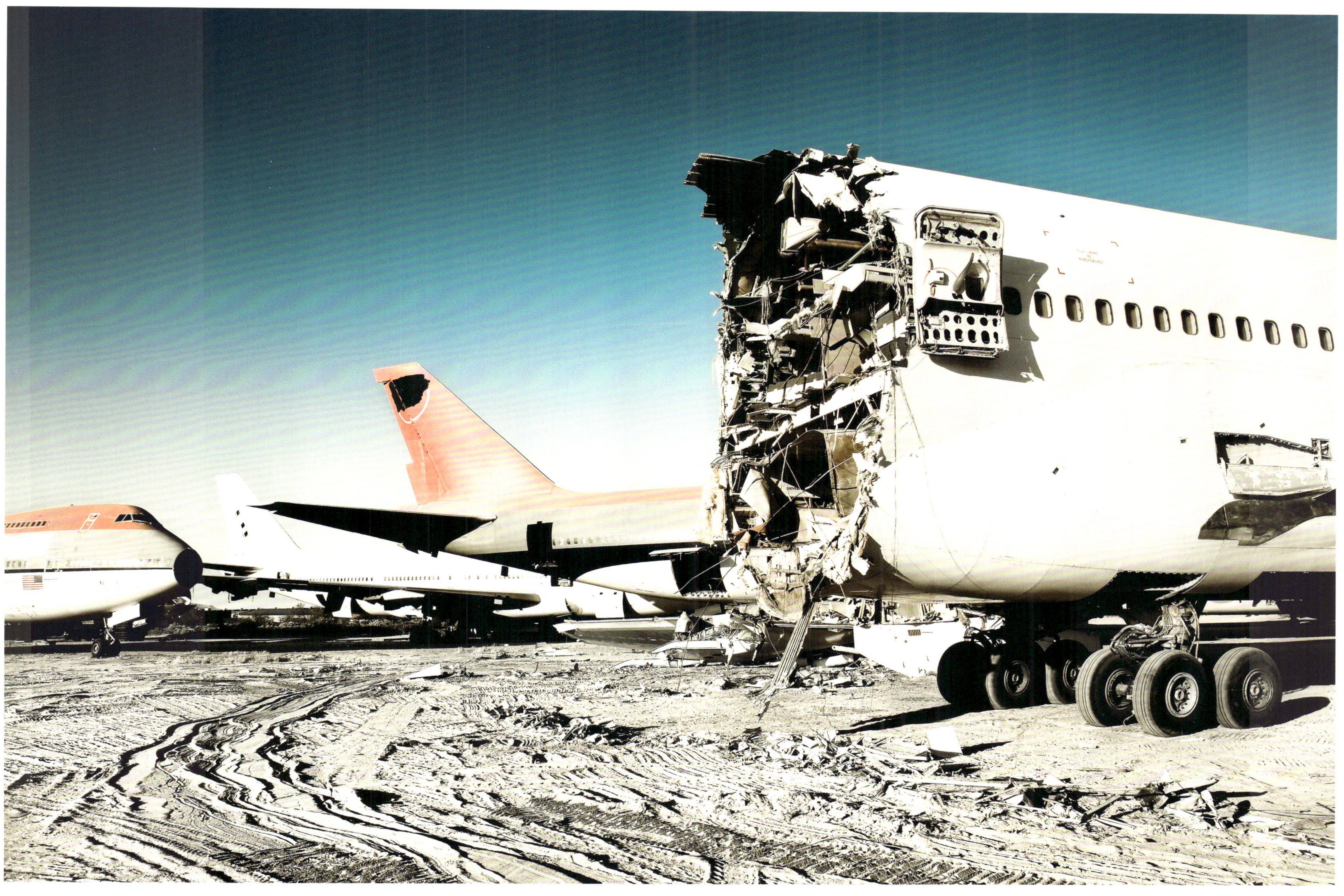

Grounded #1

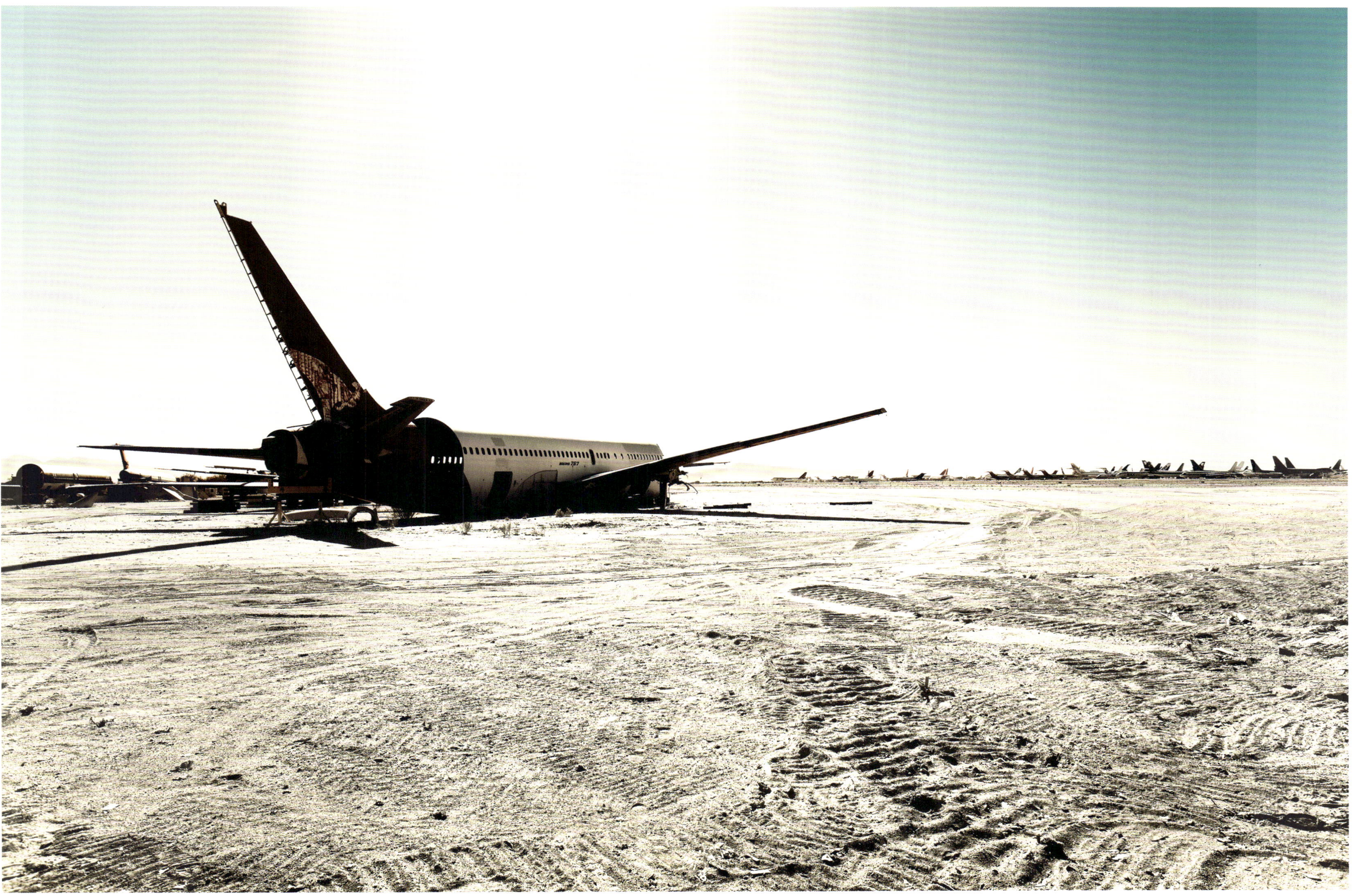

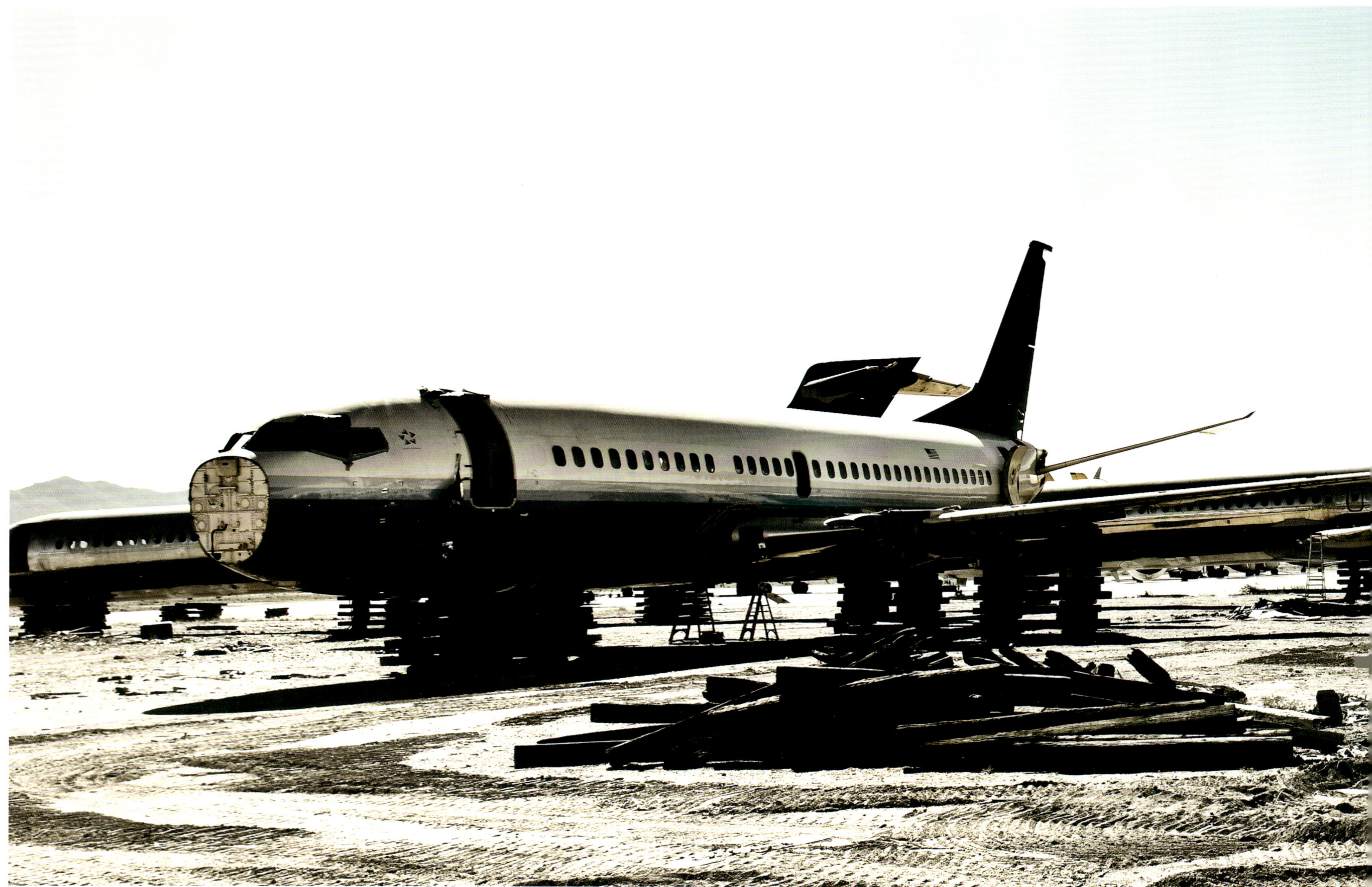

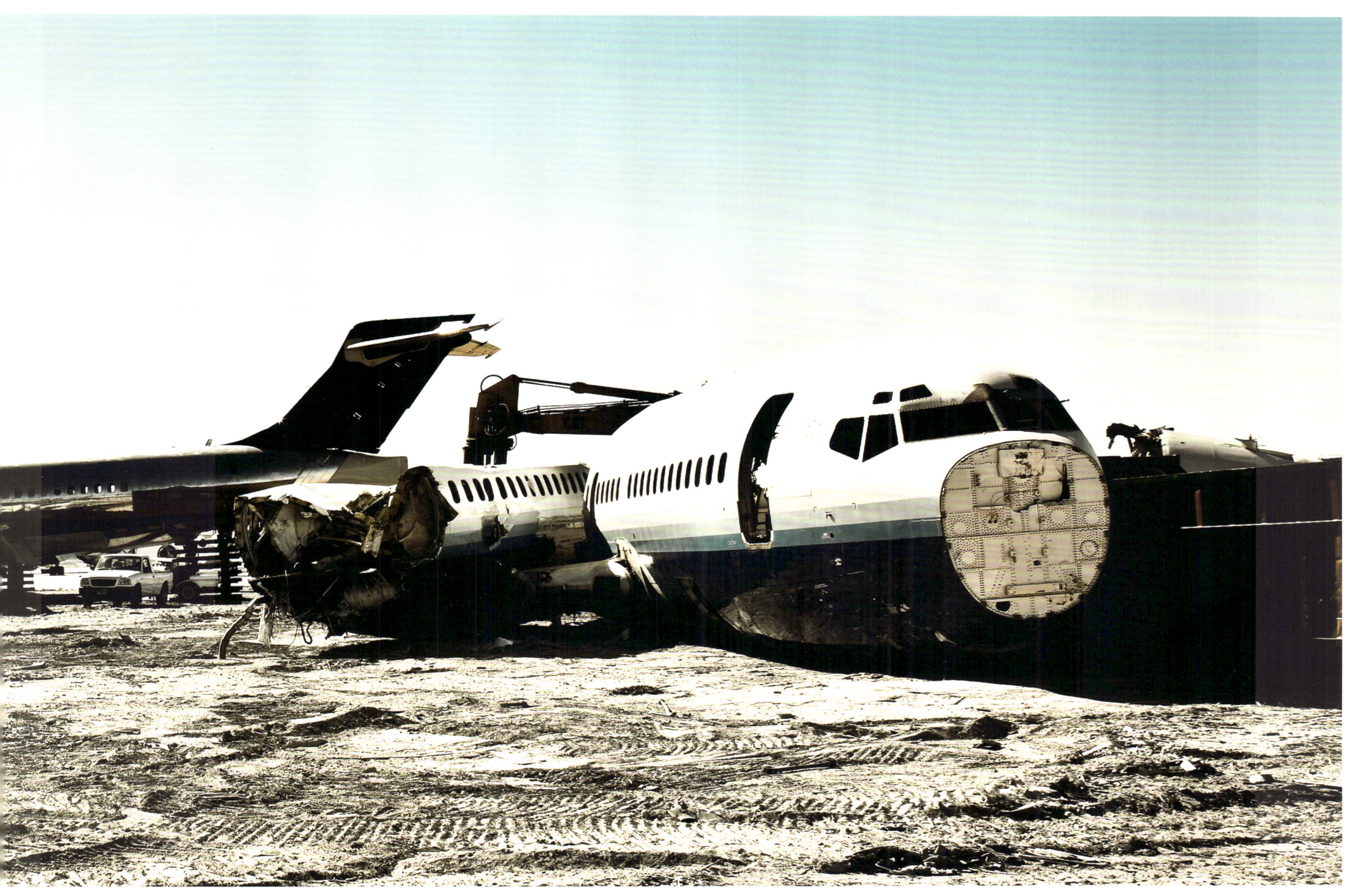

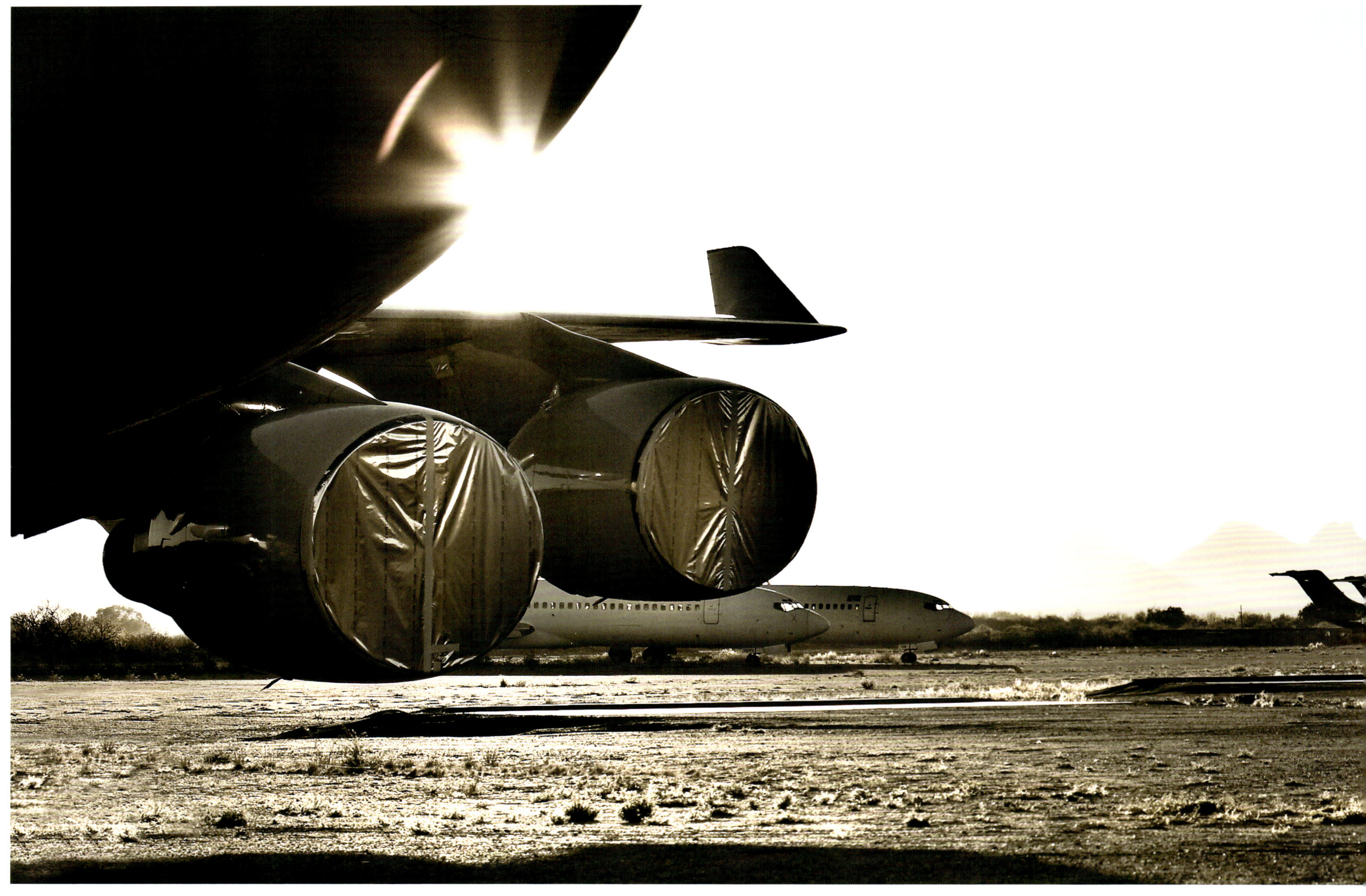

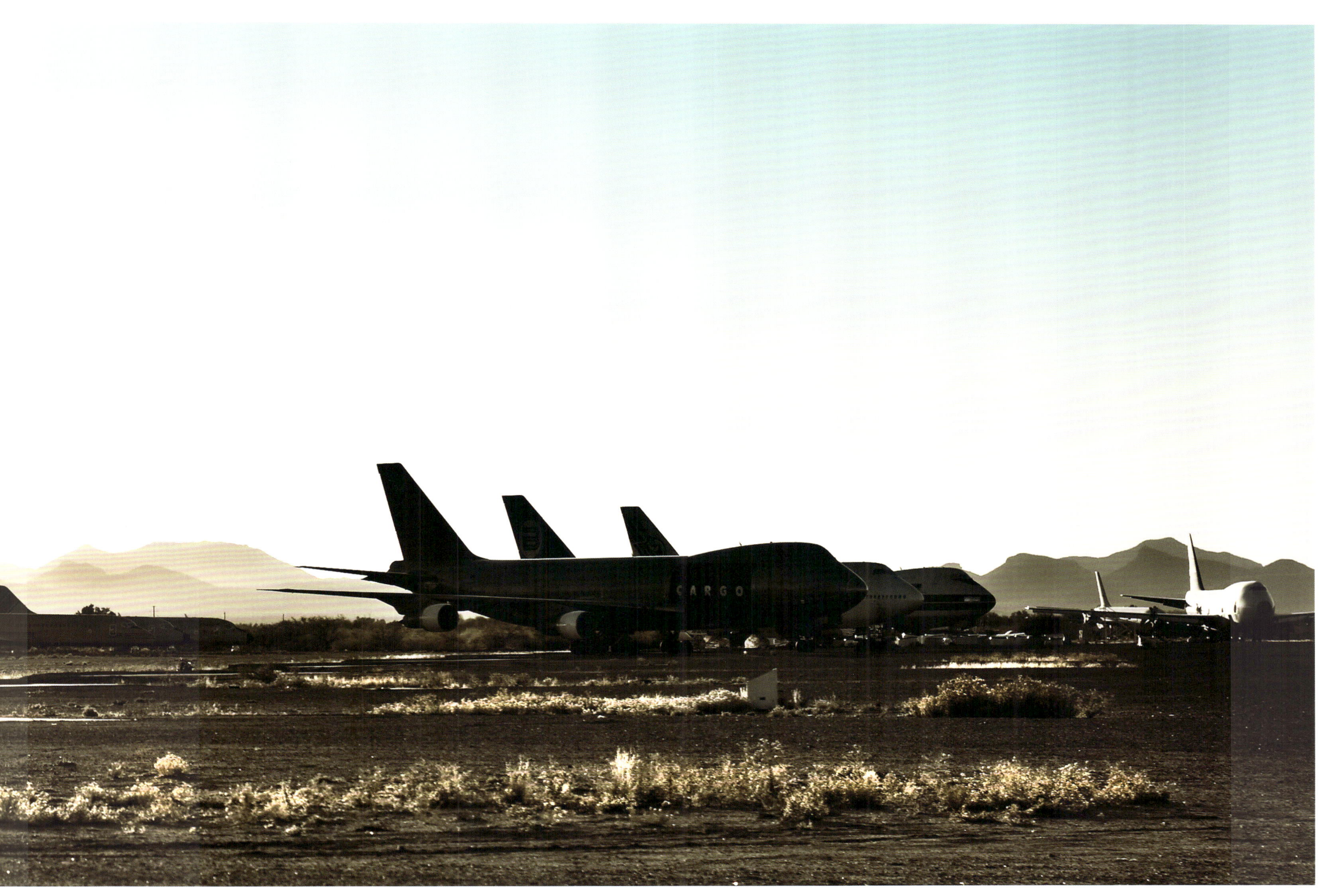

<< Boneyard #3
<< Stranded #3
Blue Sky

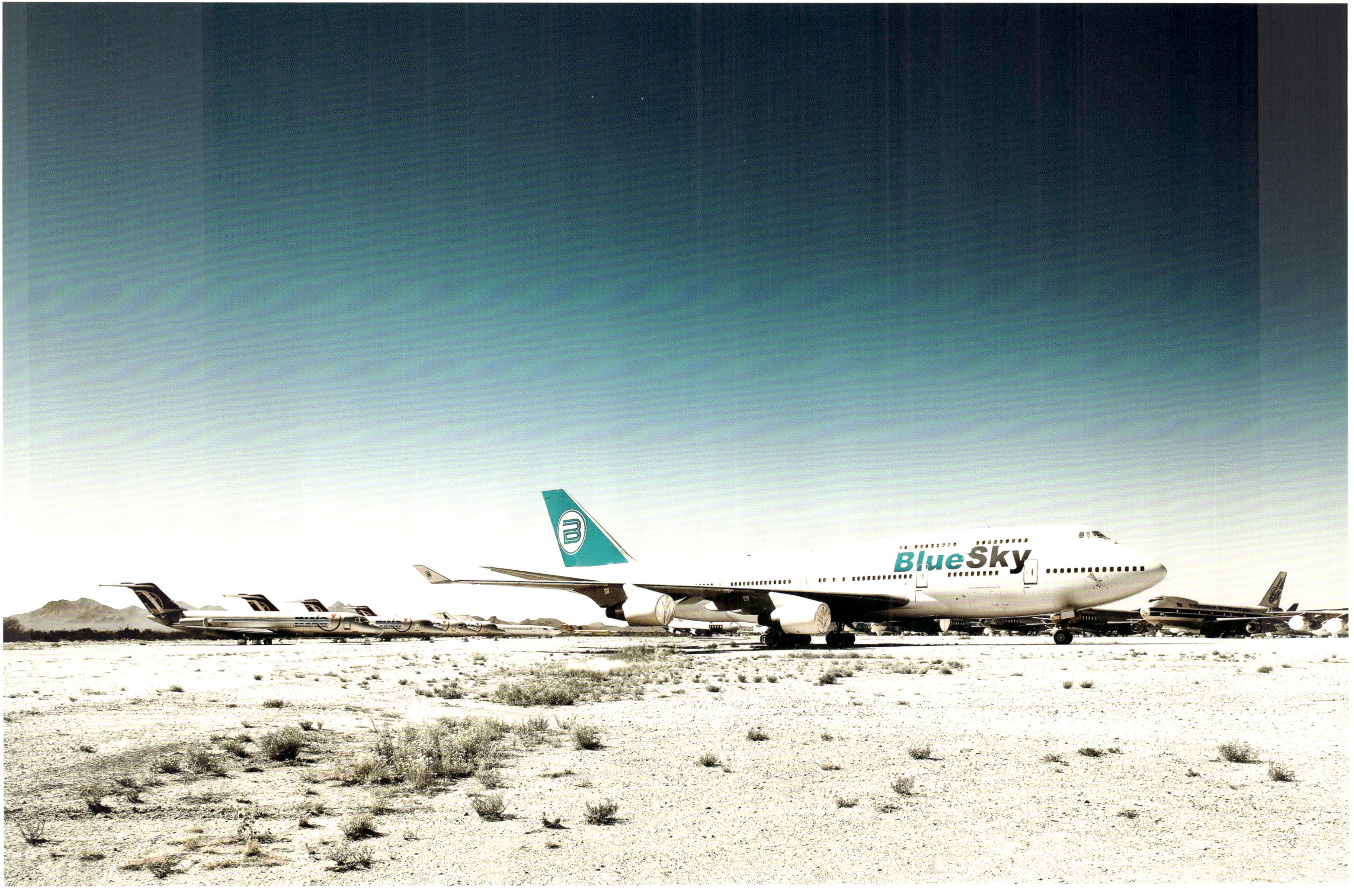
BlueSky

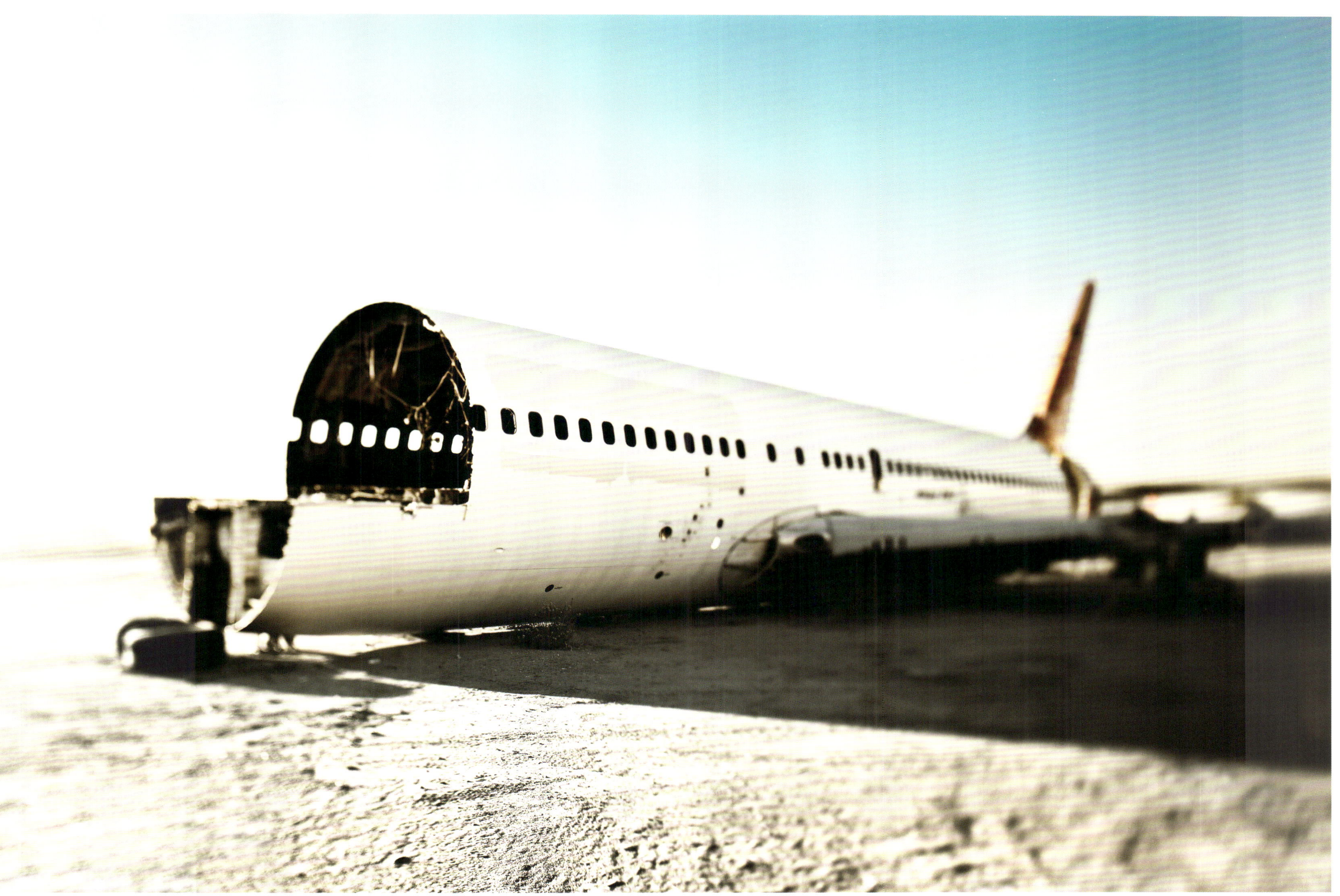

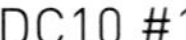

DC10 #1

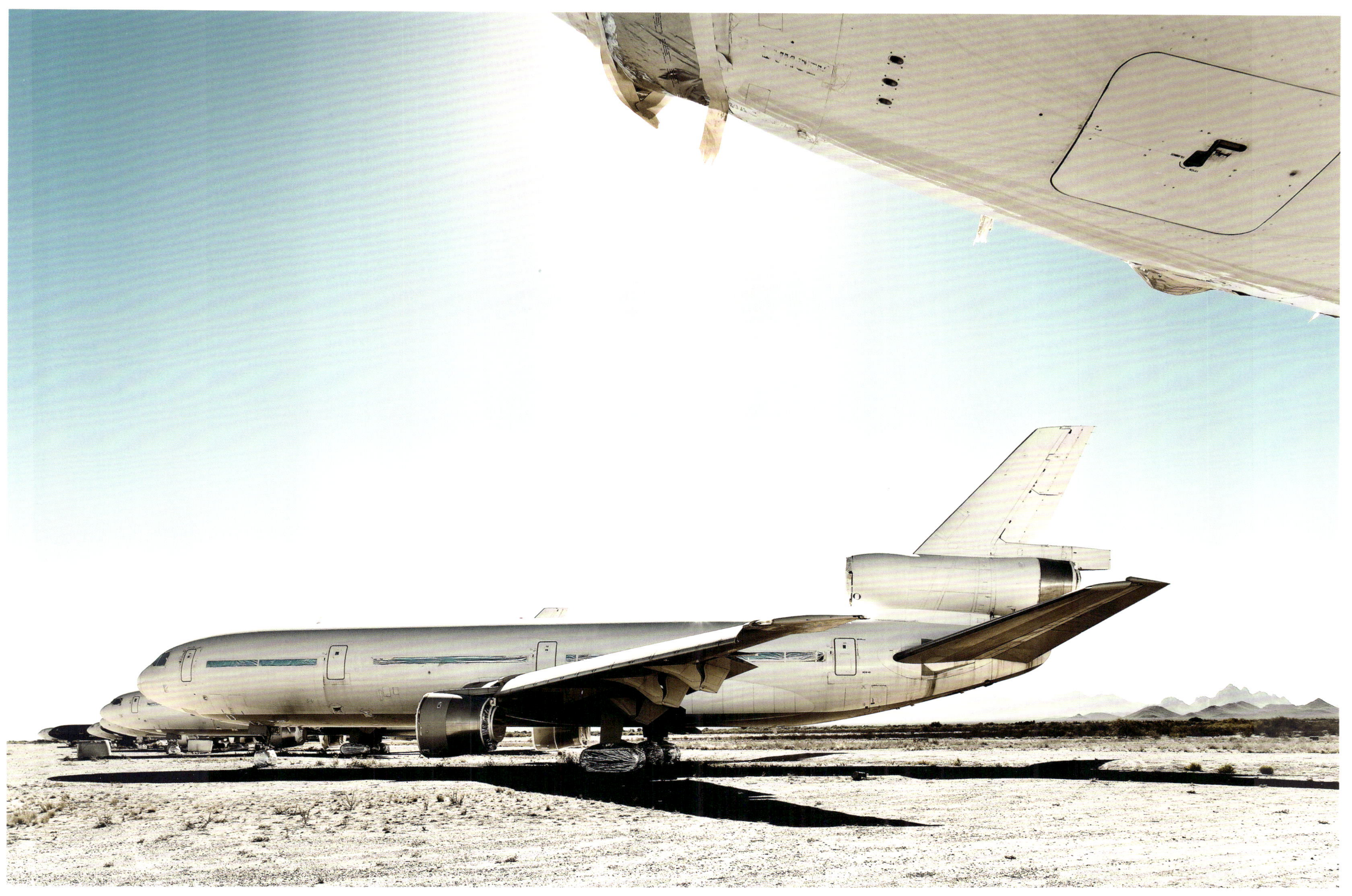

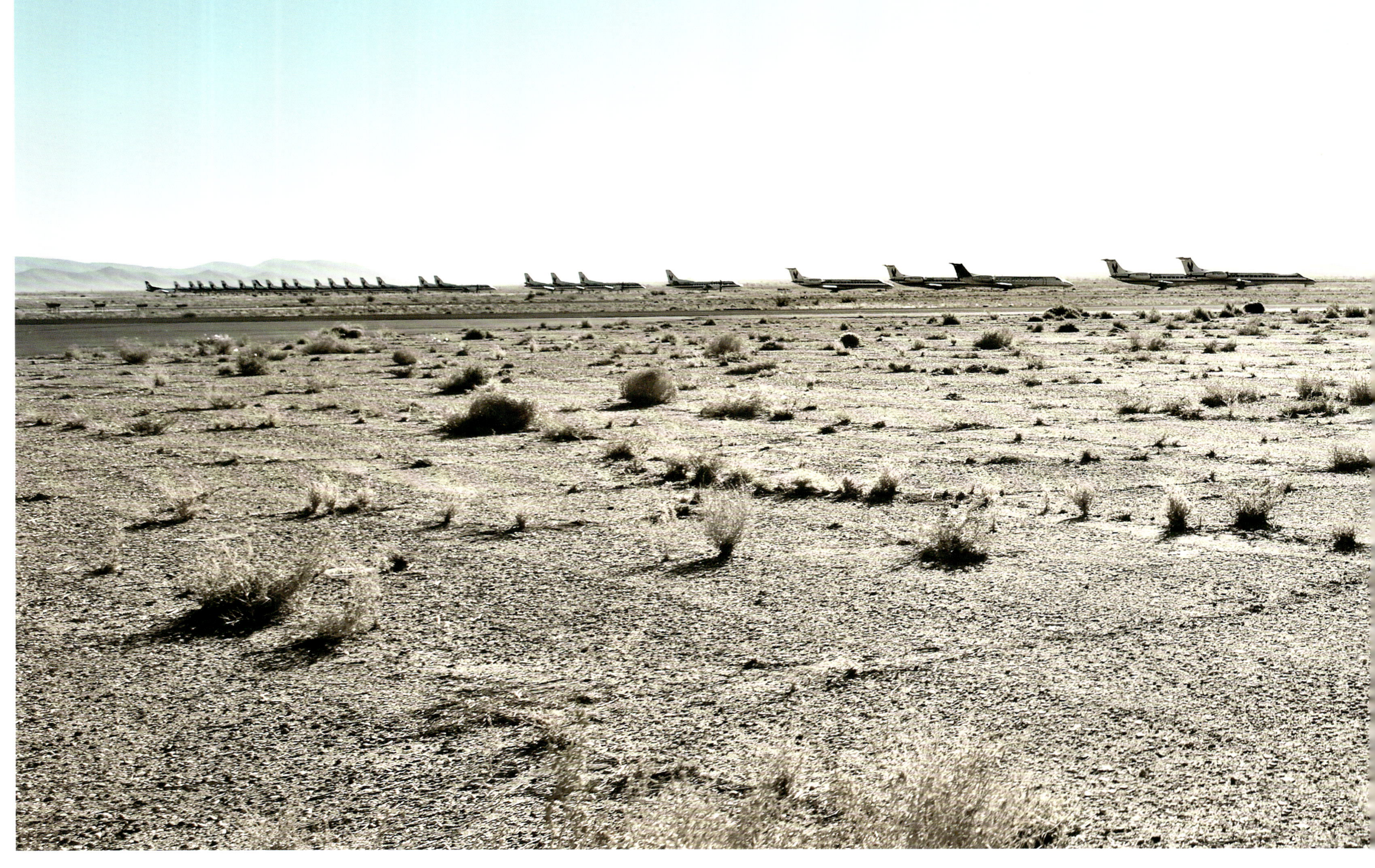

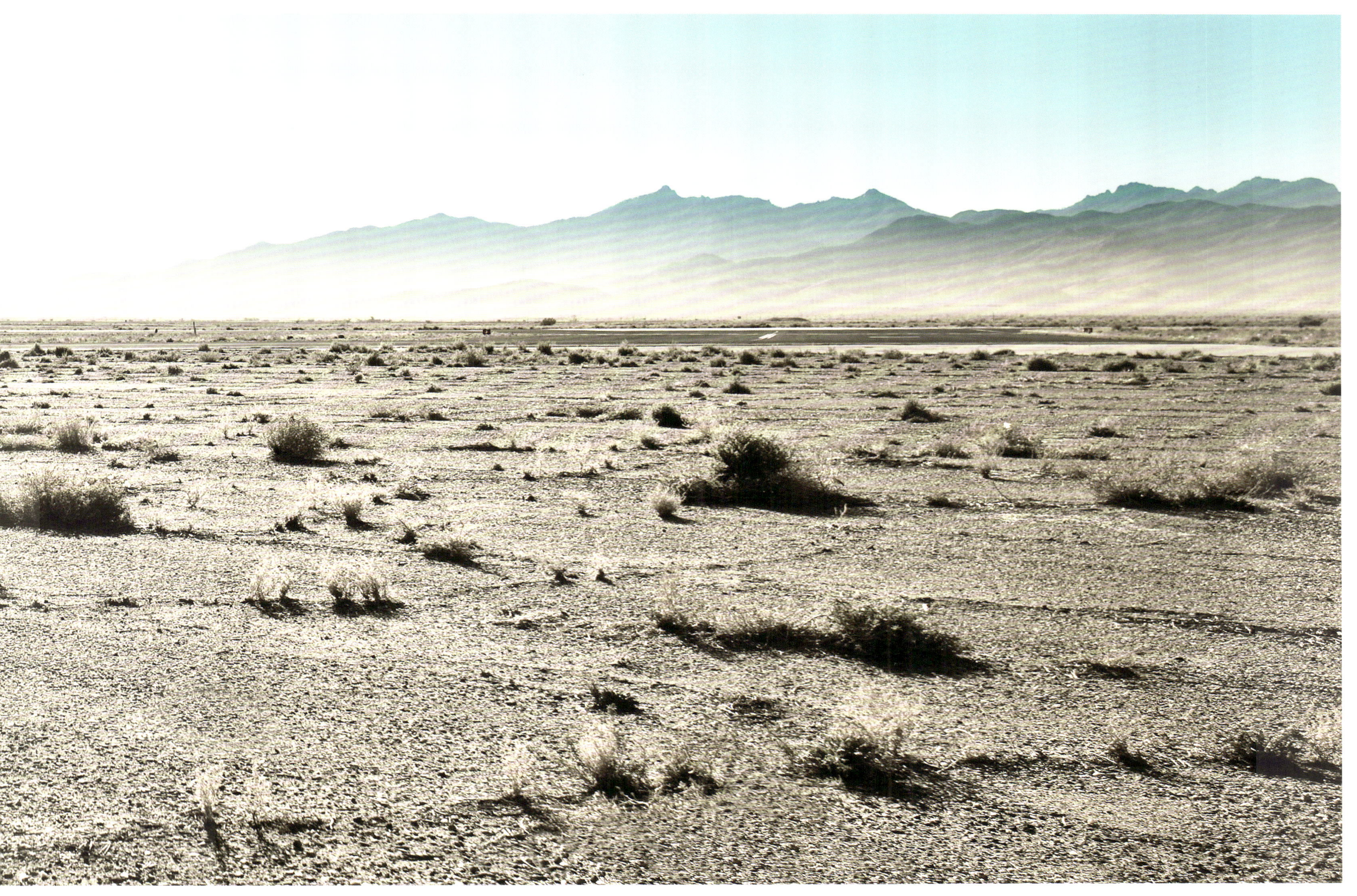

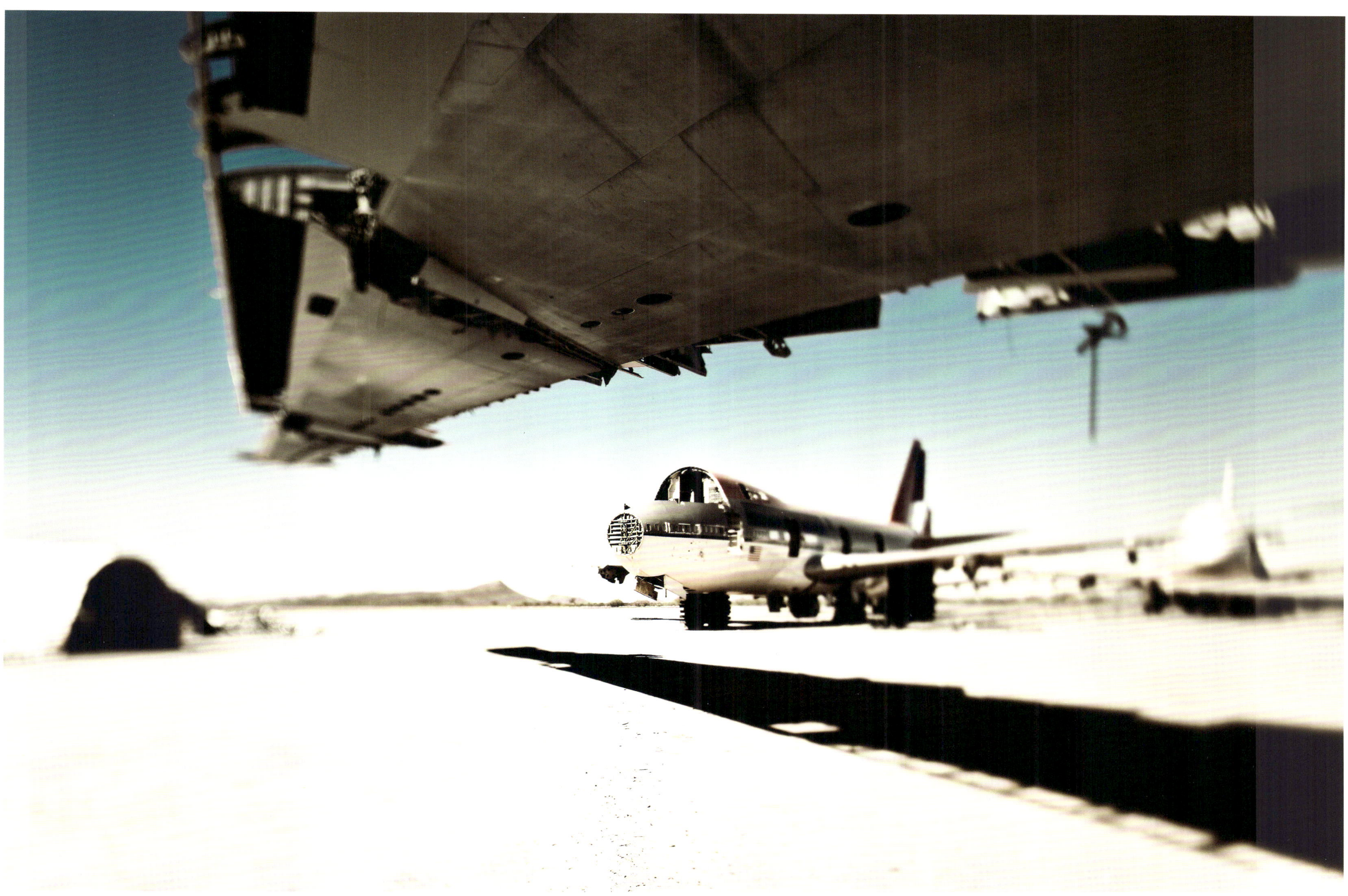

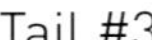

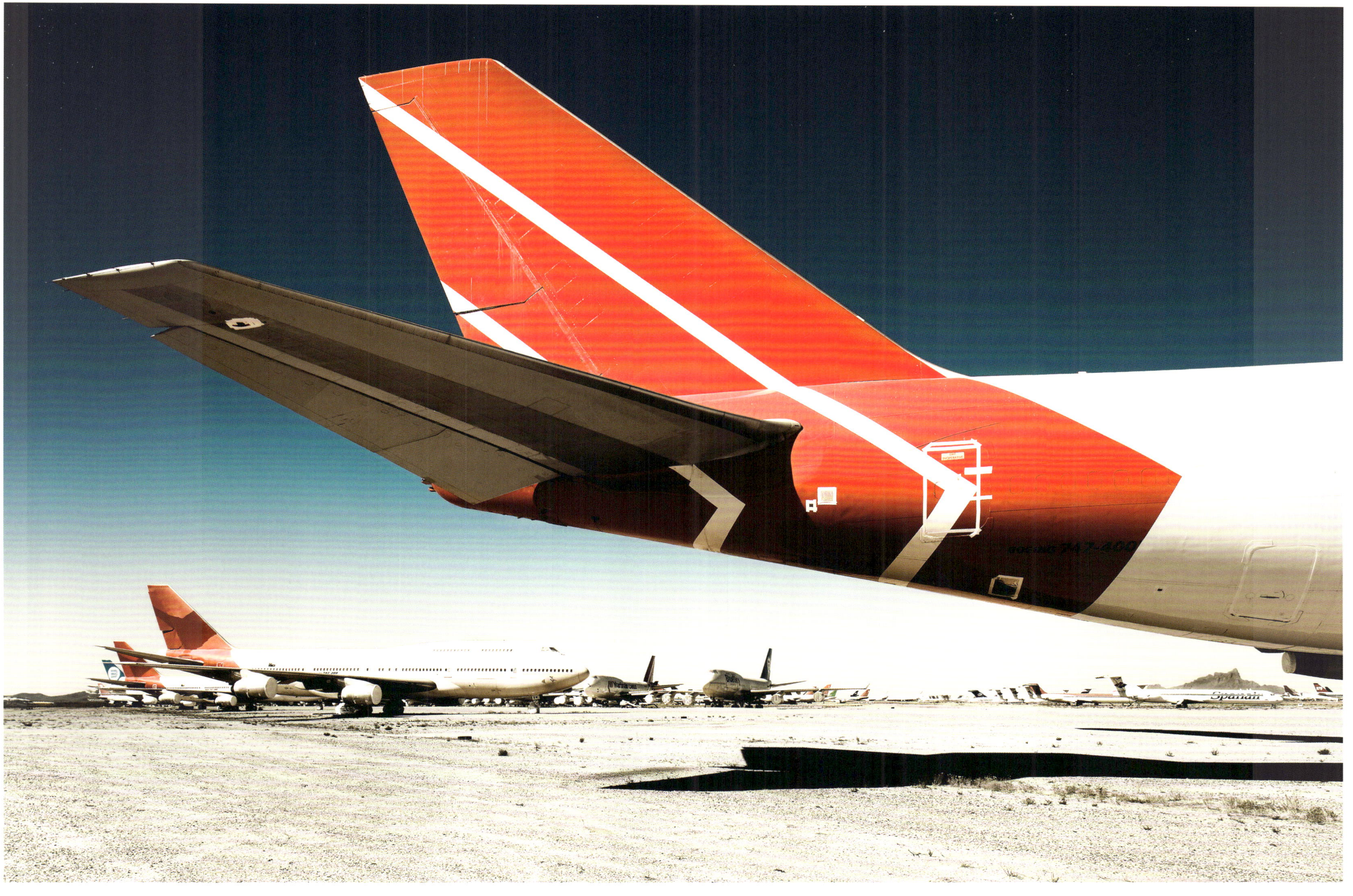

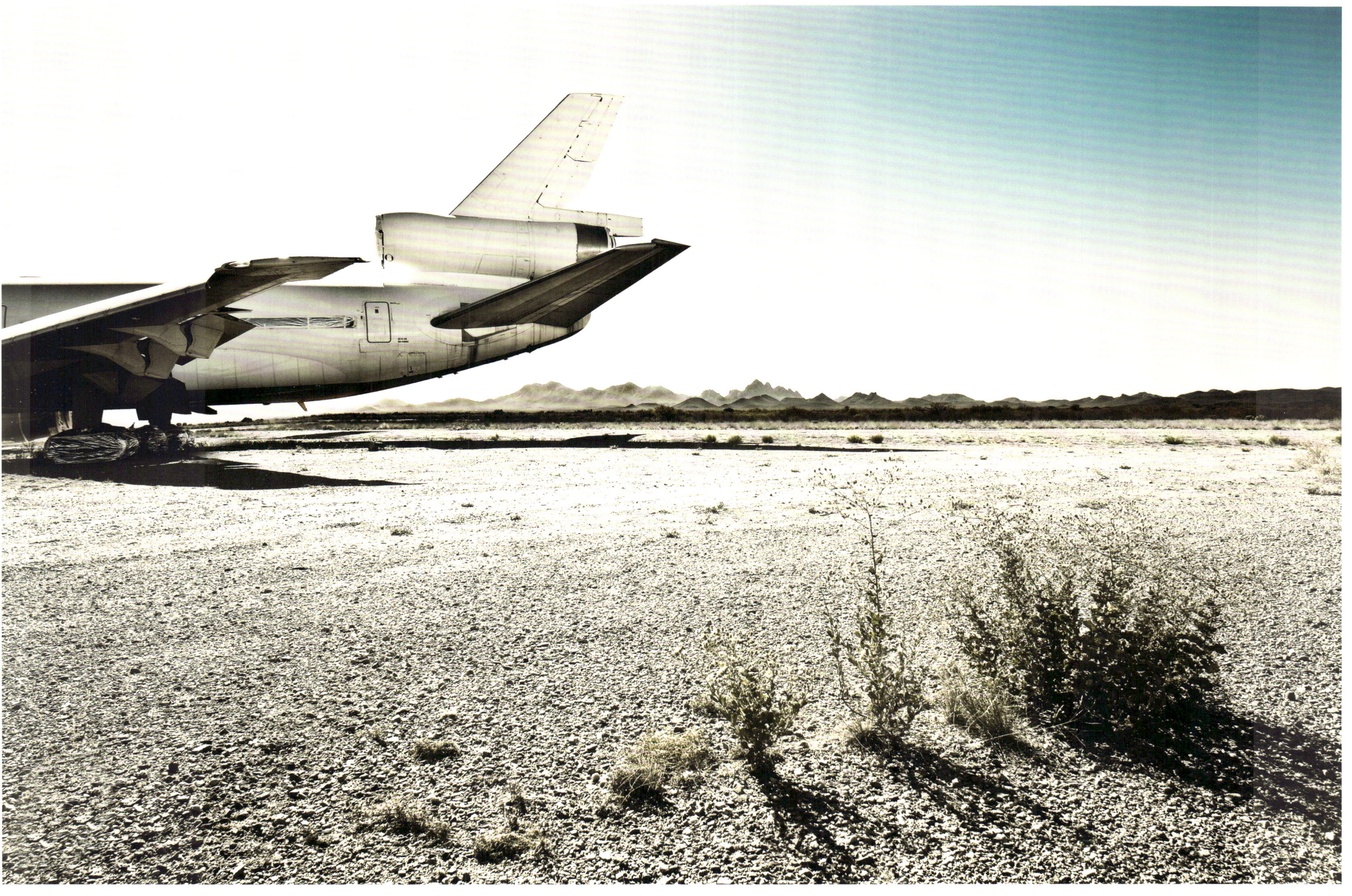

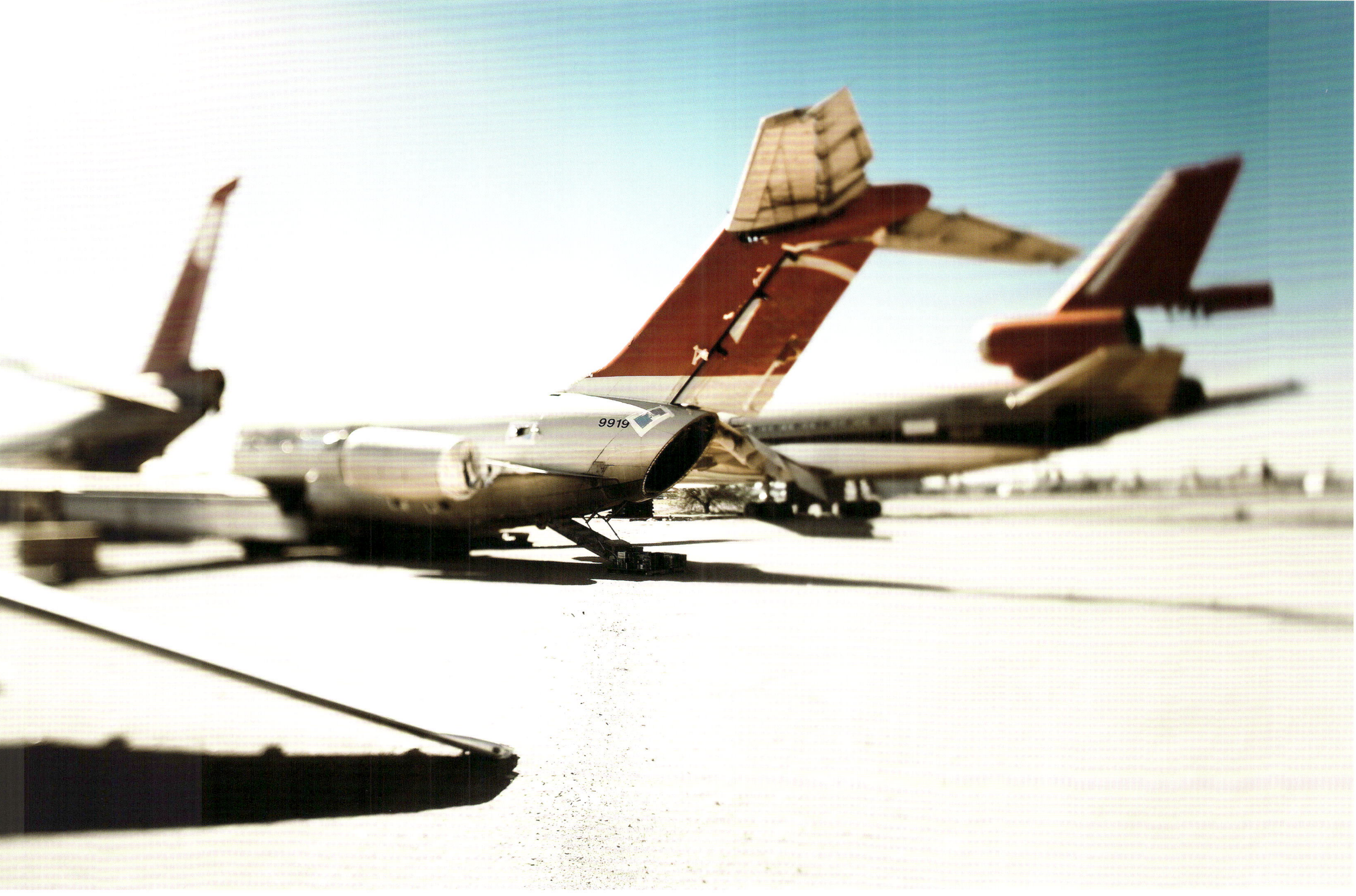

Tristar

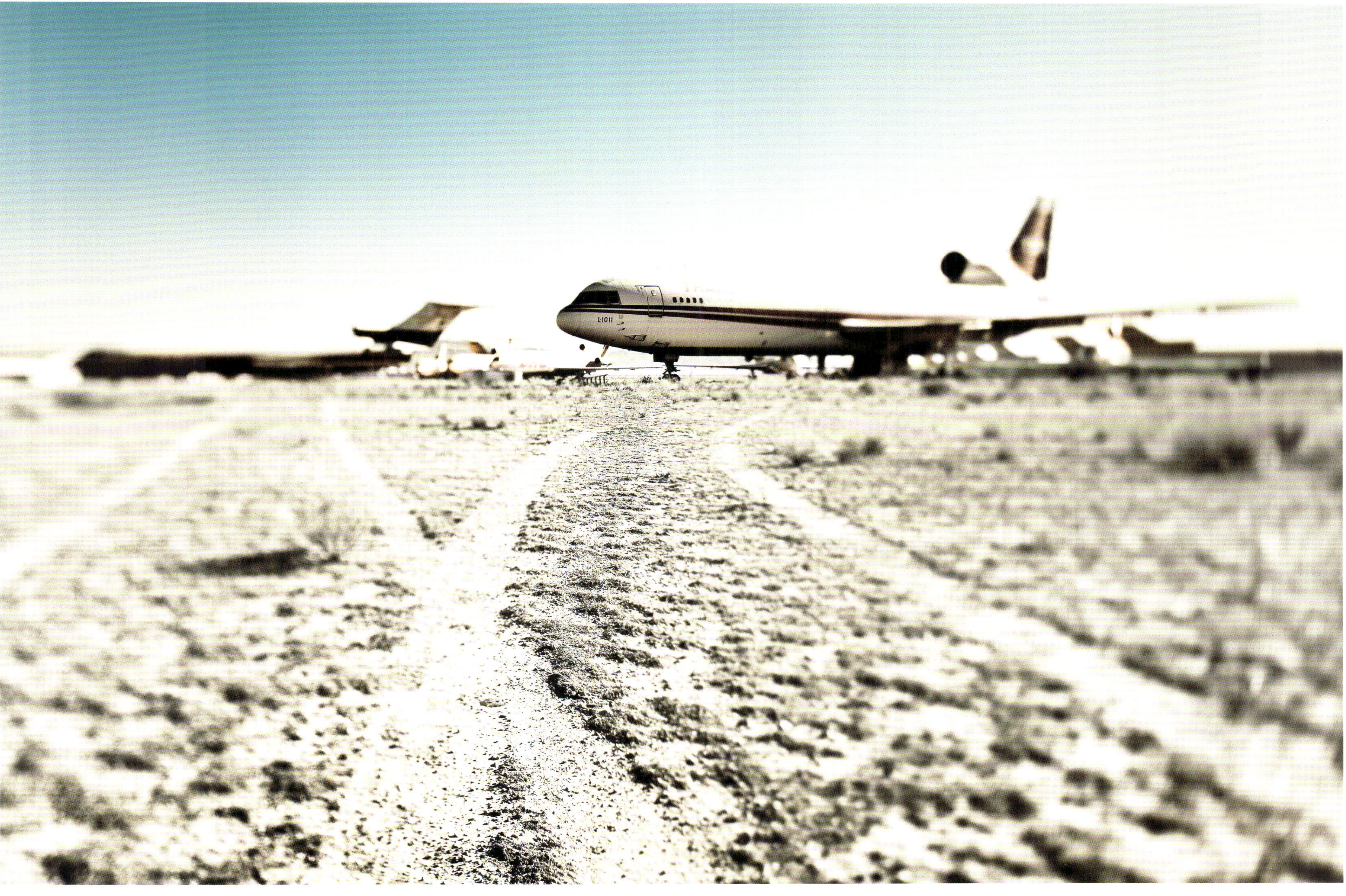

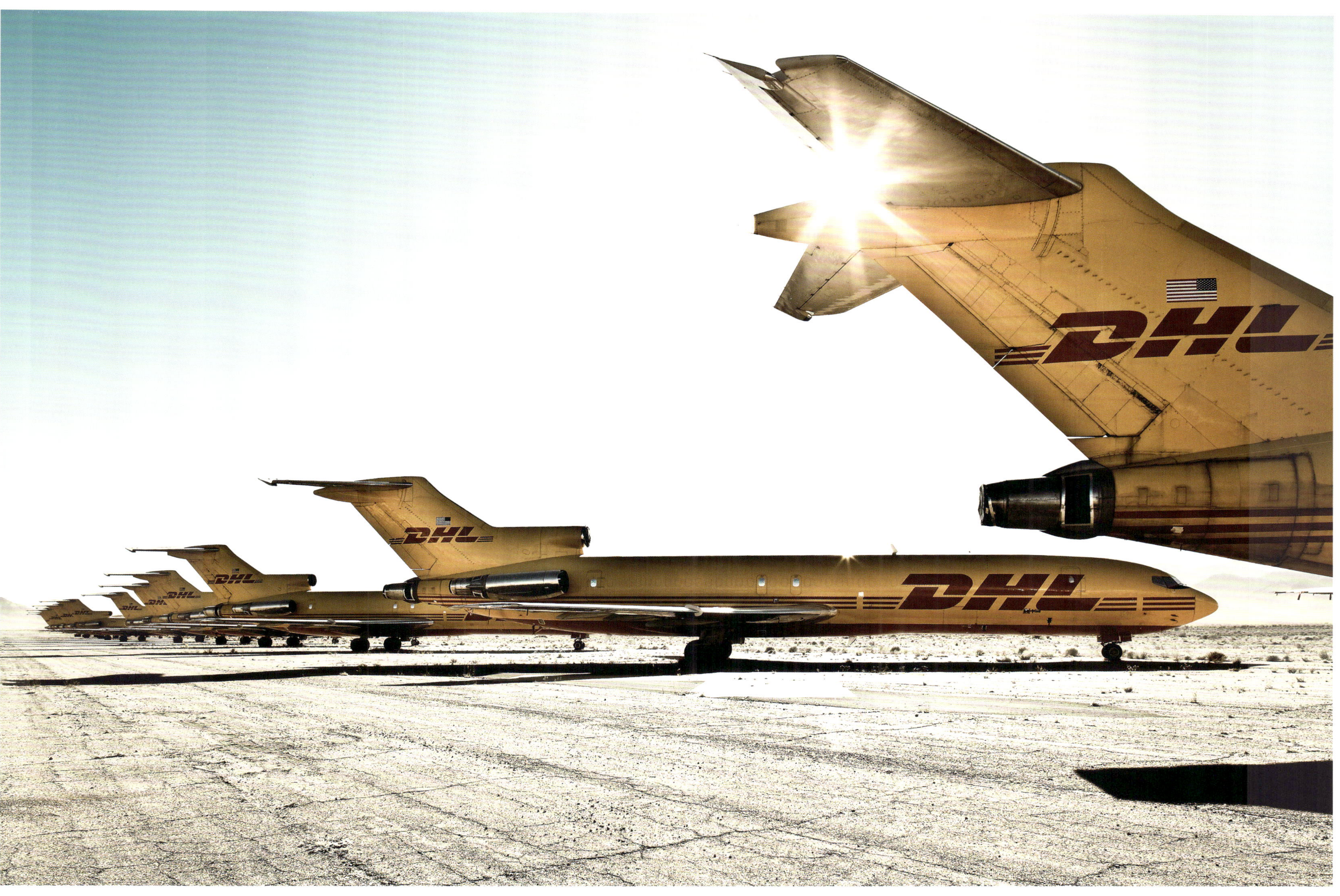

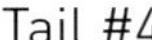

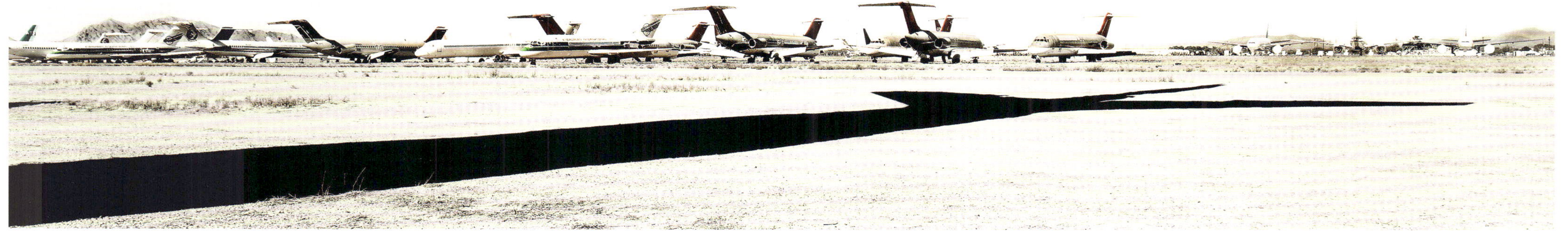

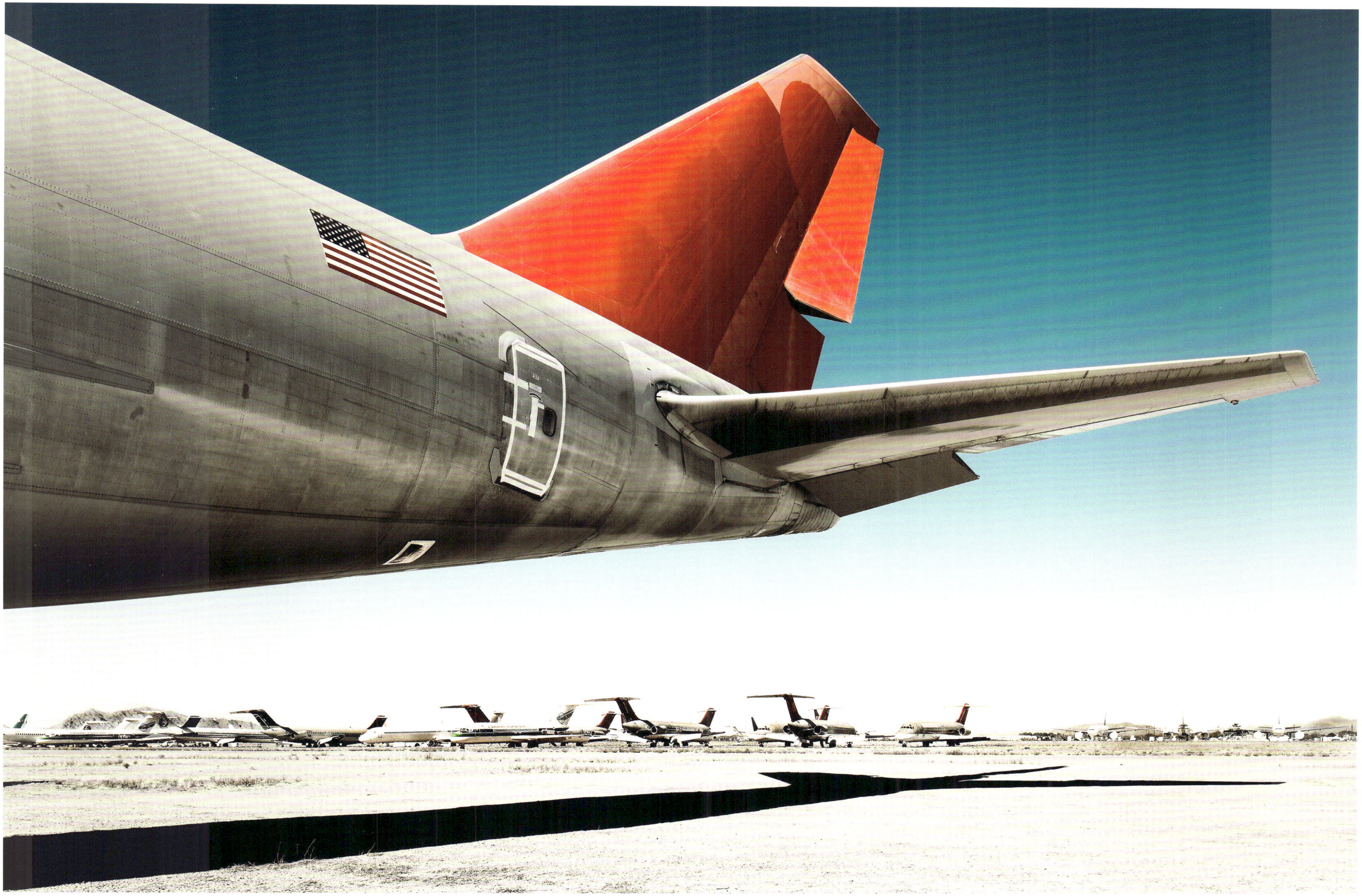

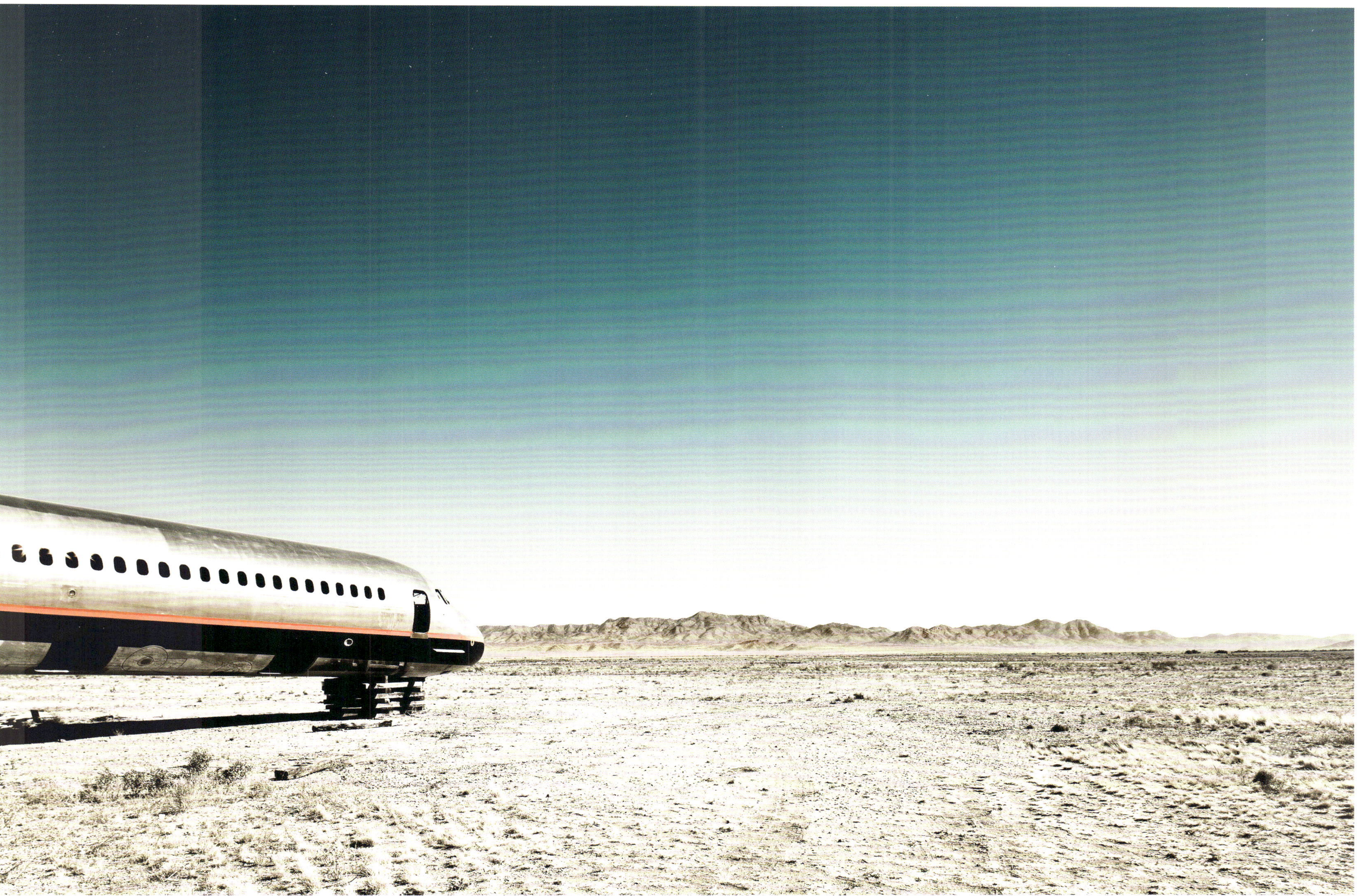

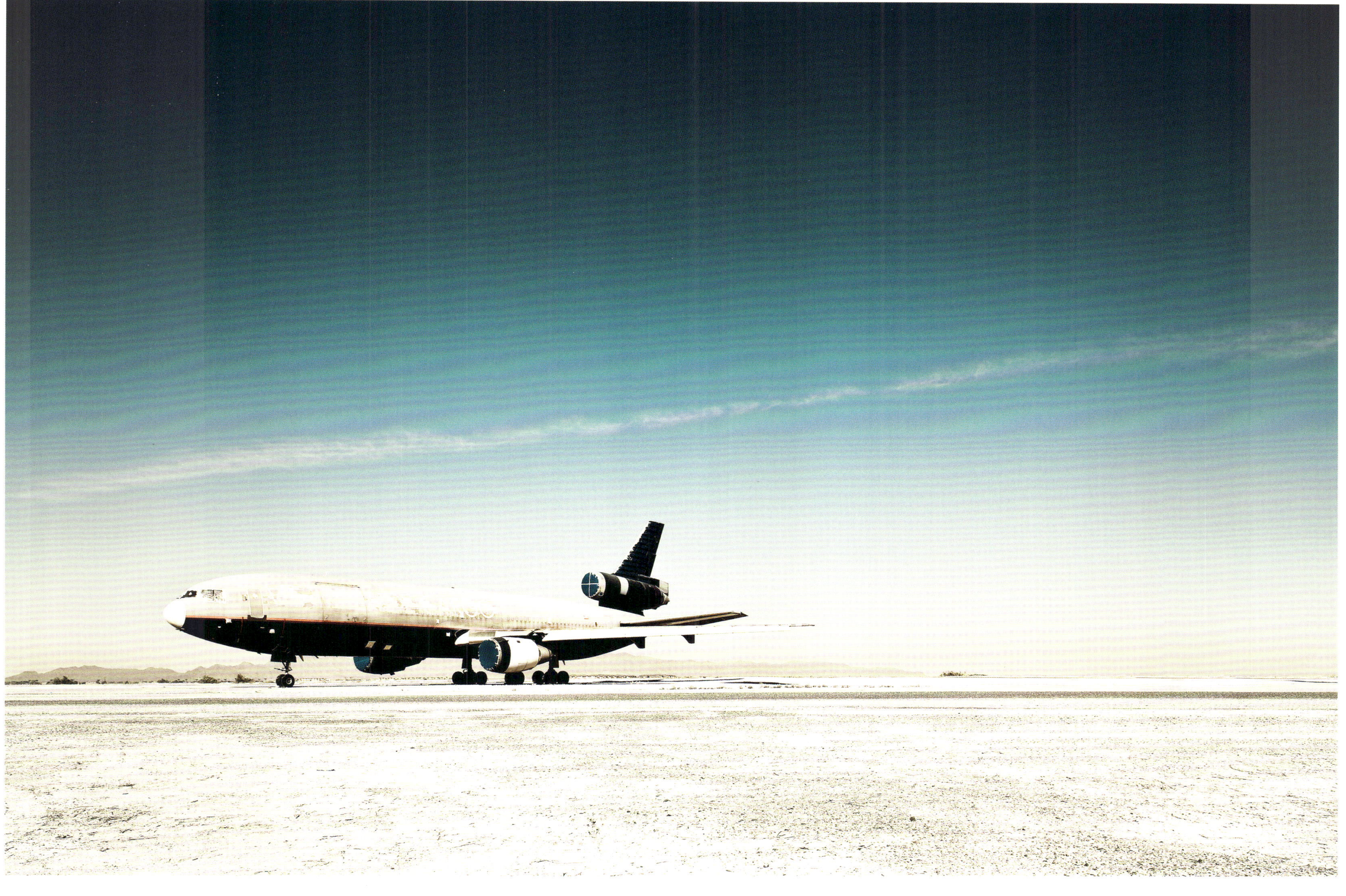

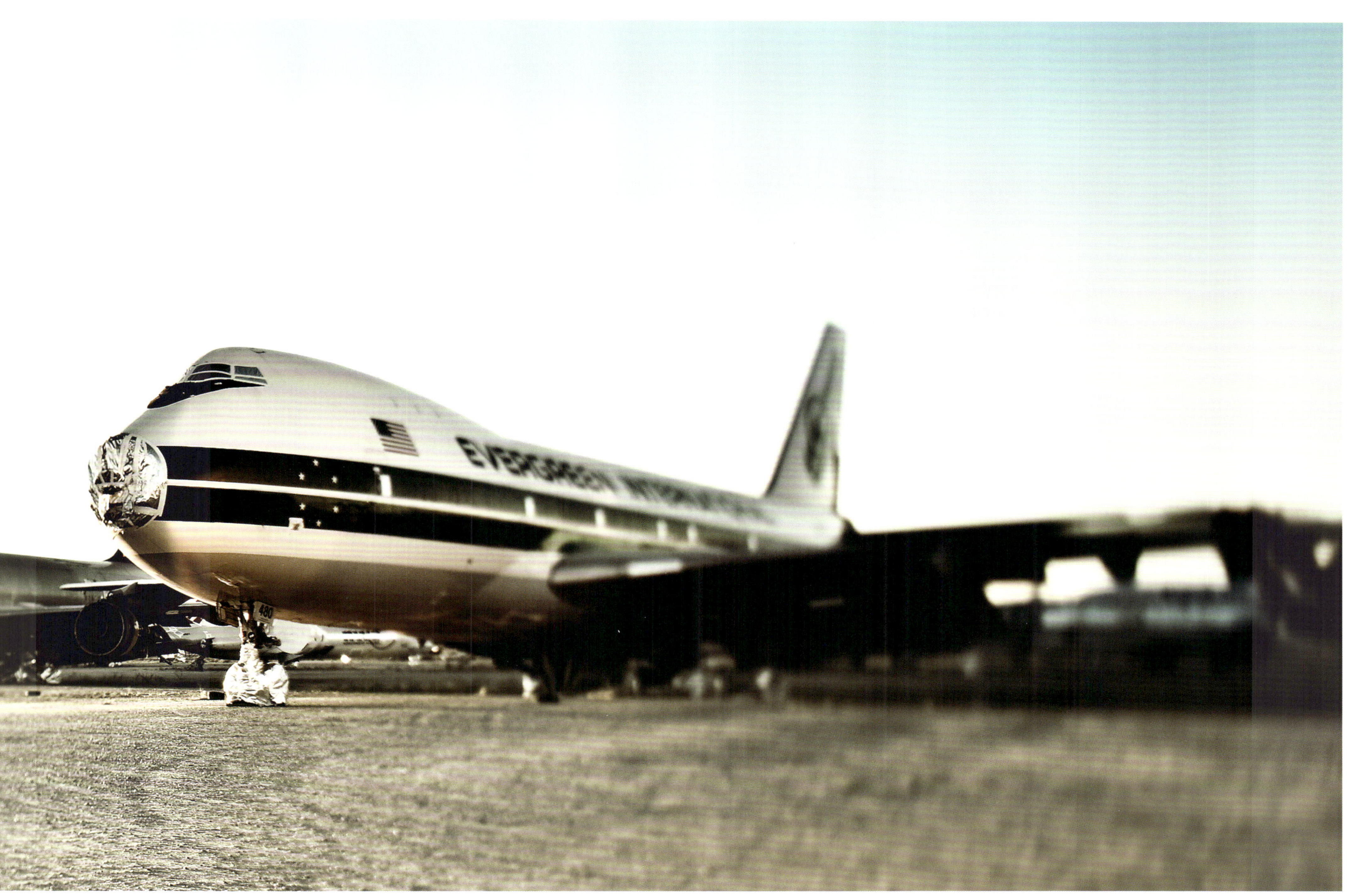
EVERGREEN INTERNATIONAL

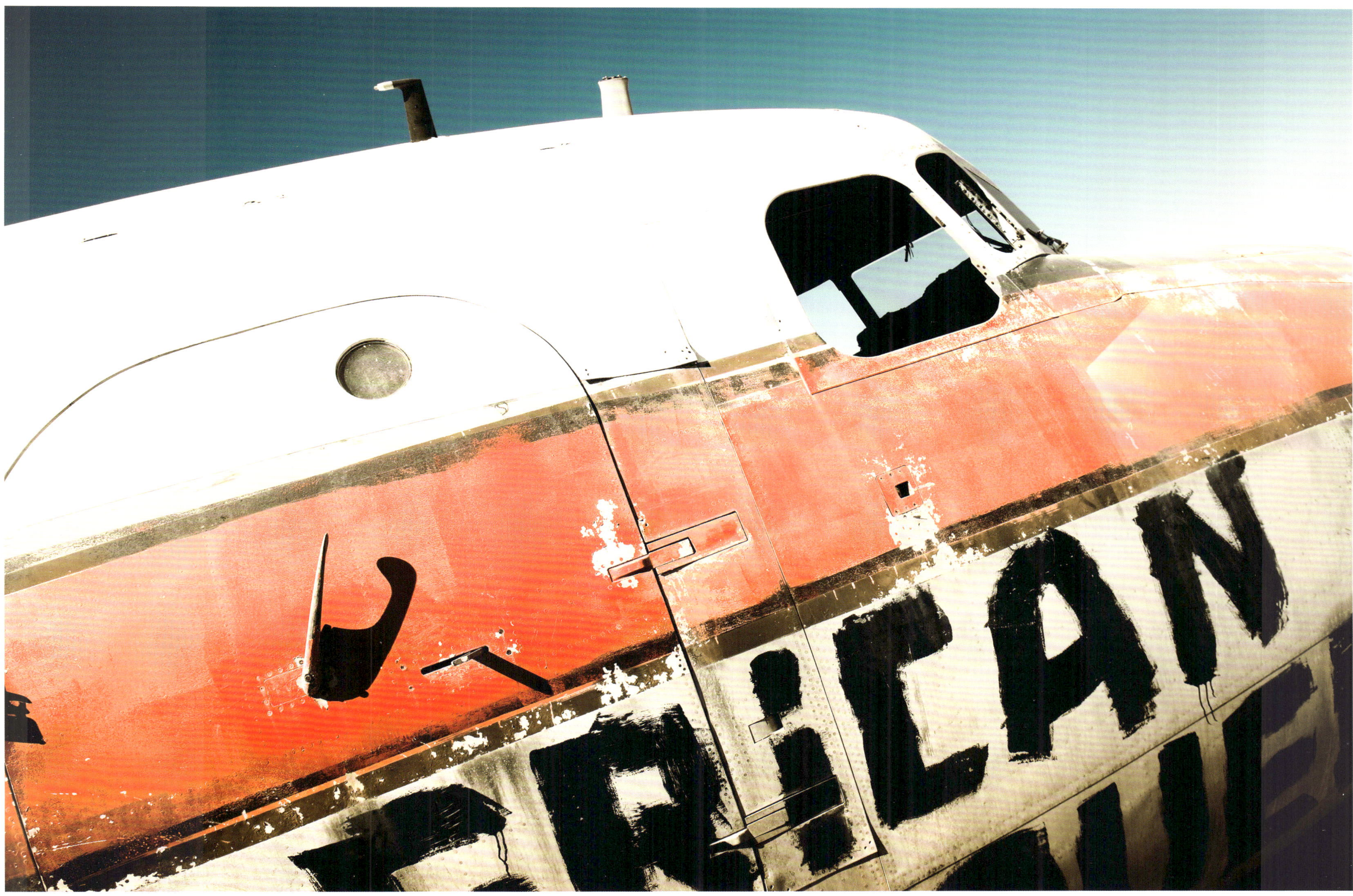

Dakotas #1

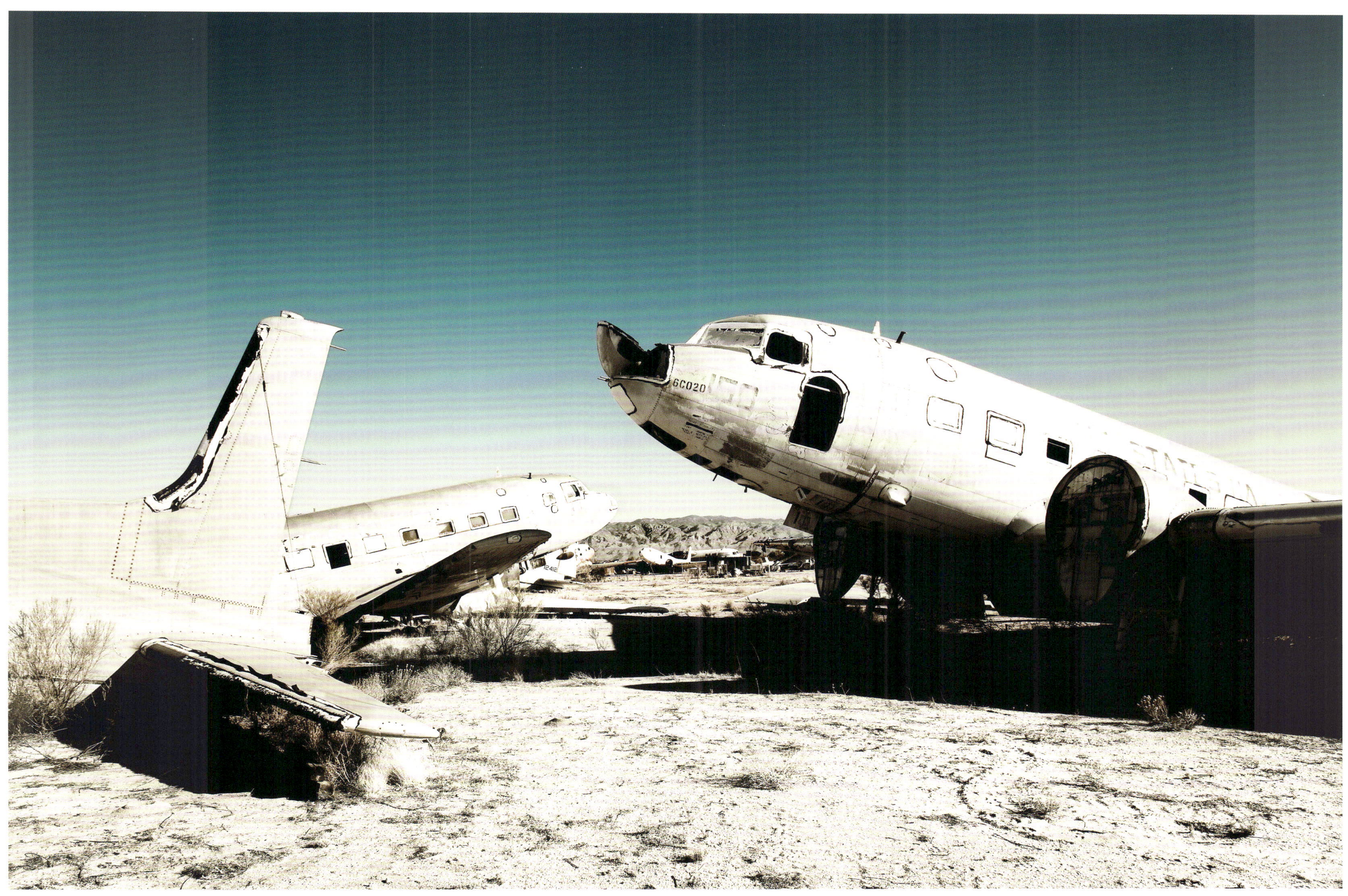

Dakotas #2

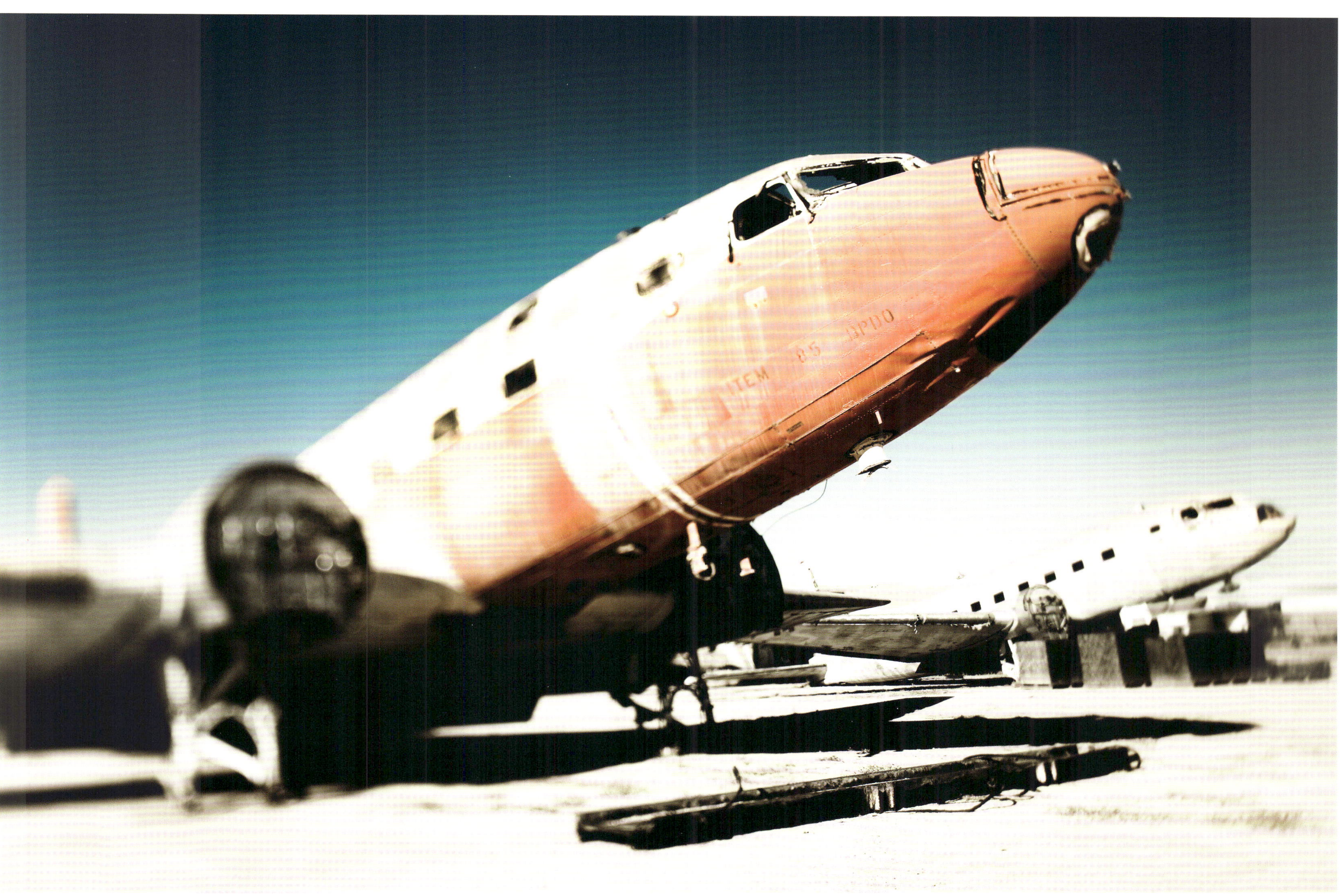

Stratocruiser

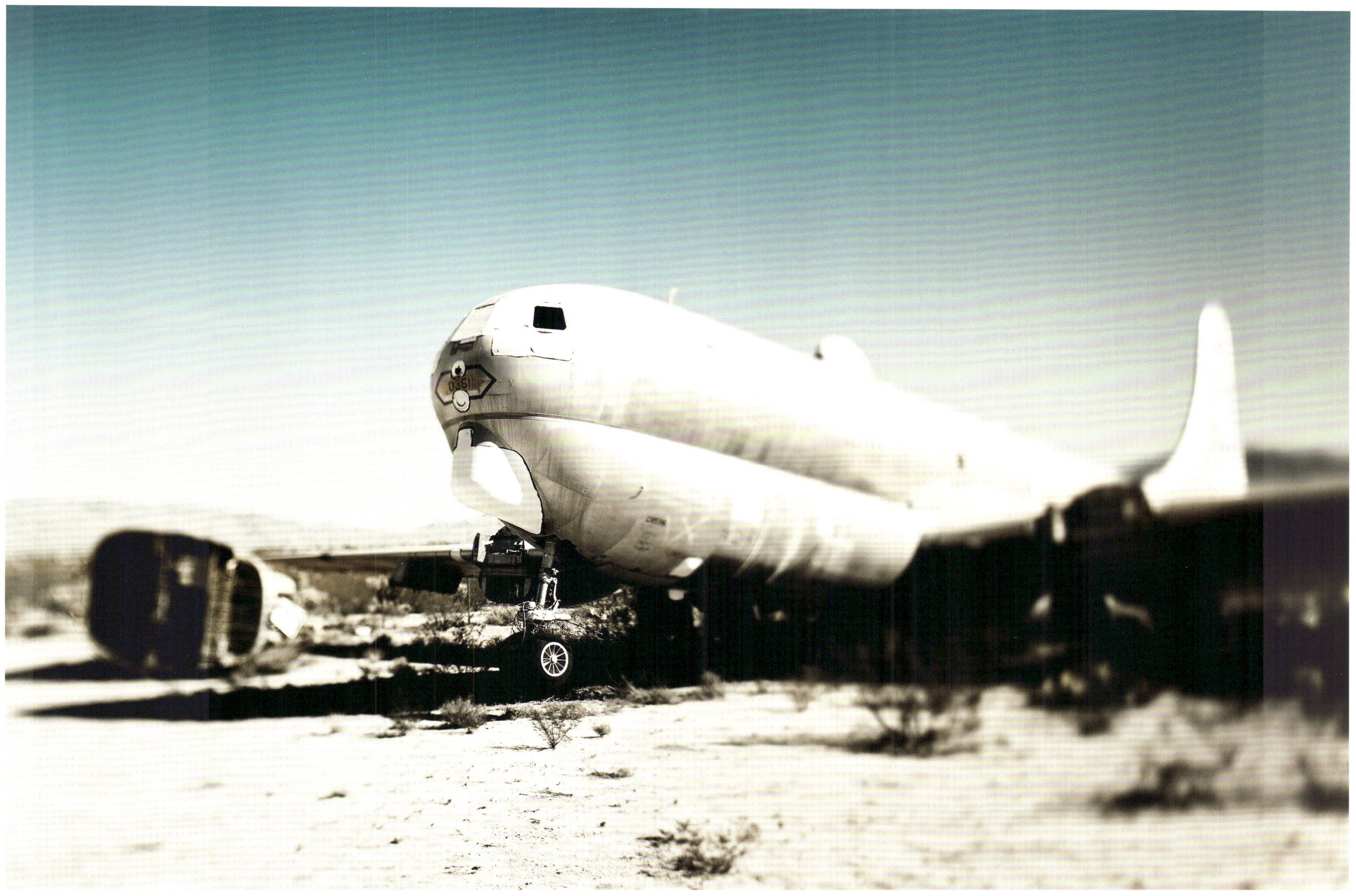

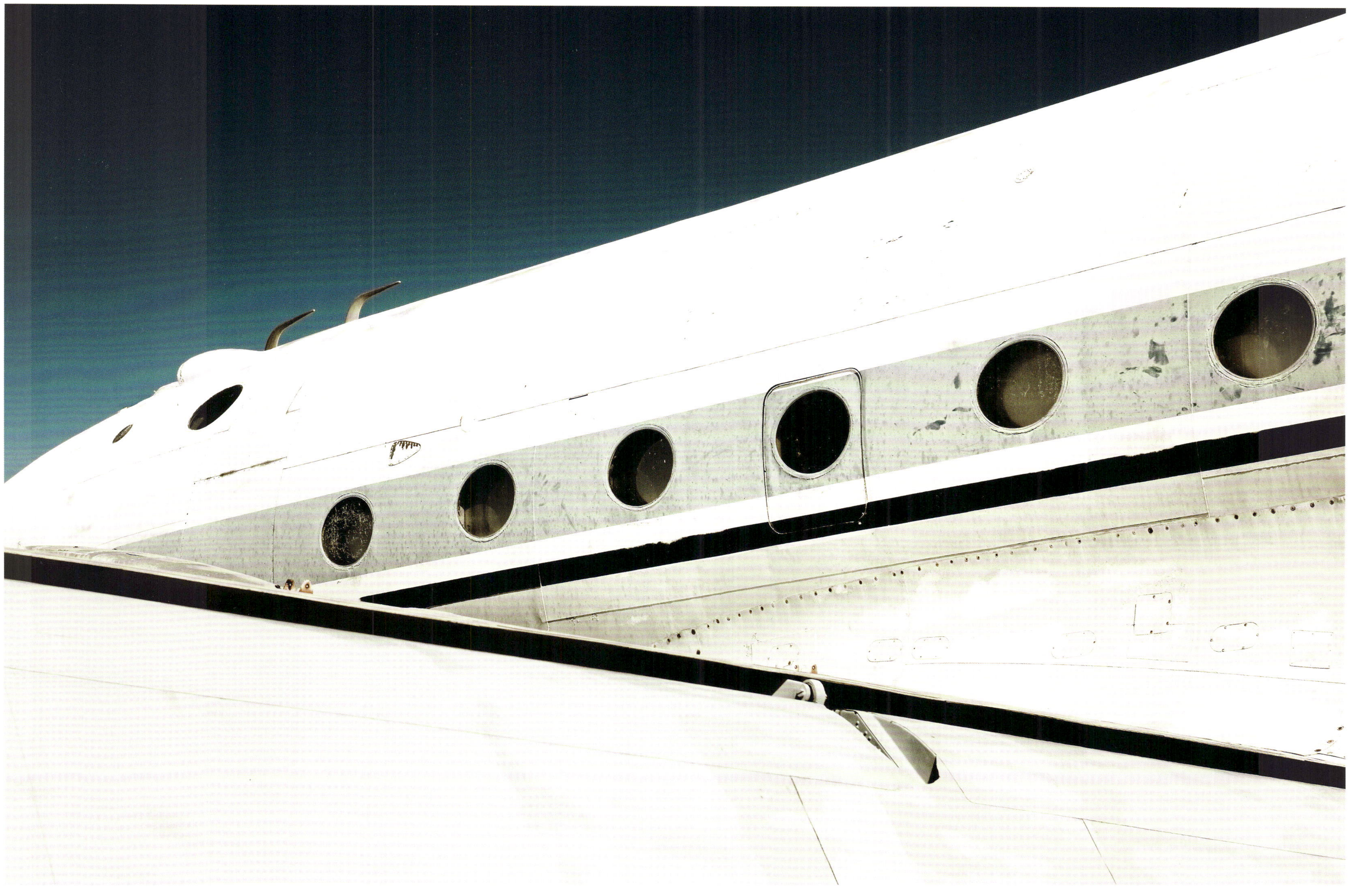

Friends

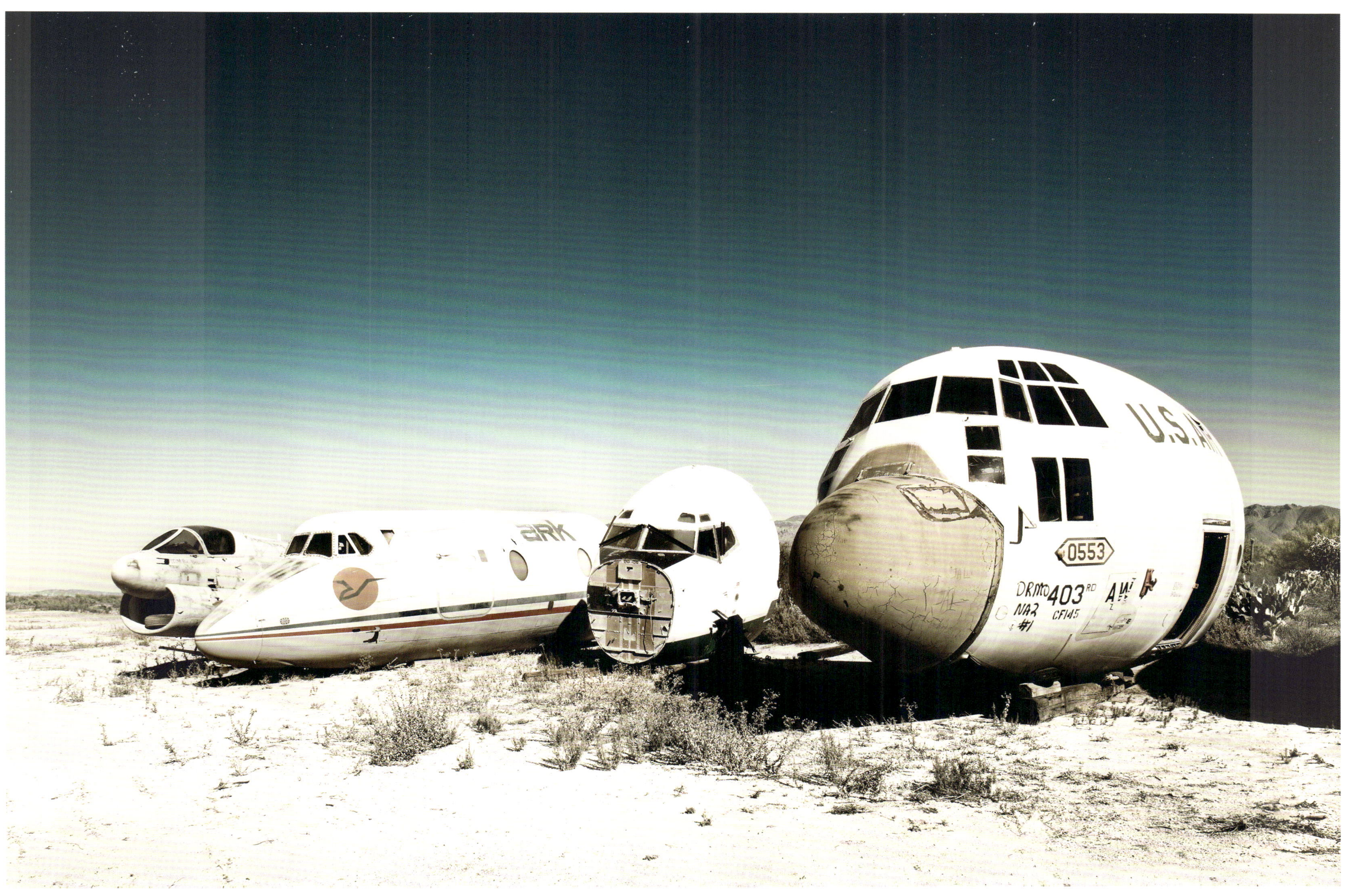
ARK
U.S.A
0553
DRMO 403 RD
NA2 CF145
#1

Dakotas #3

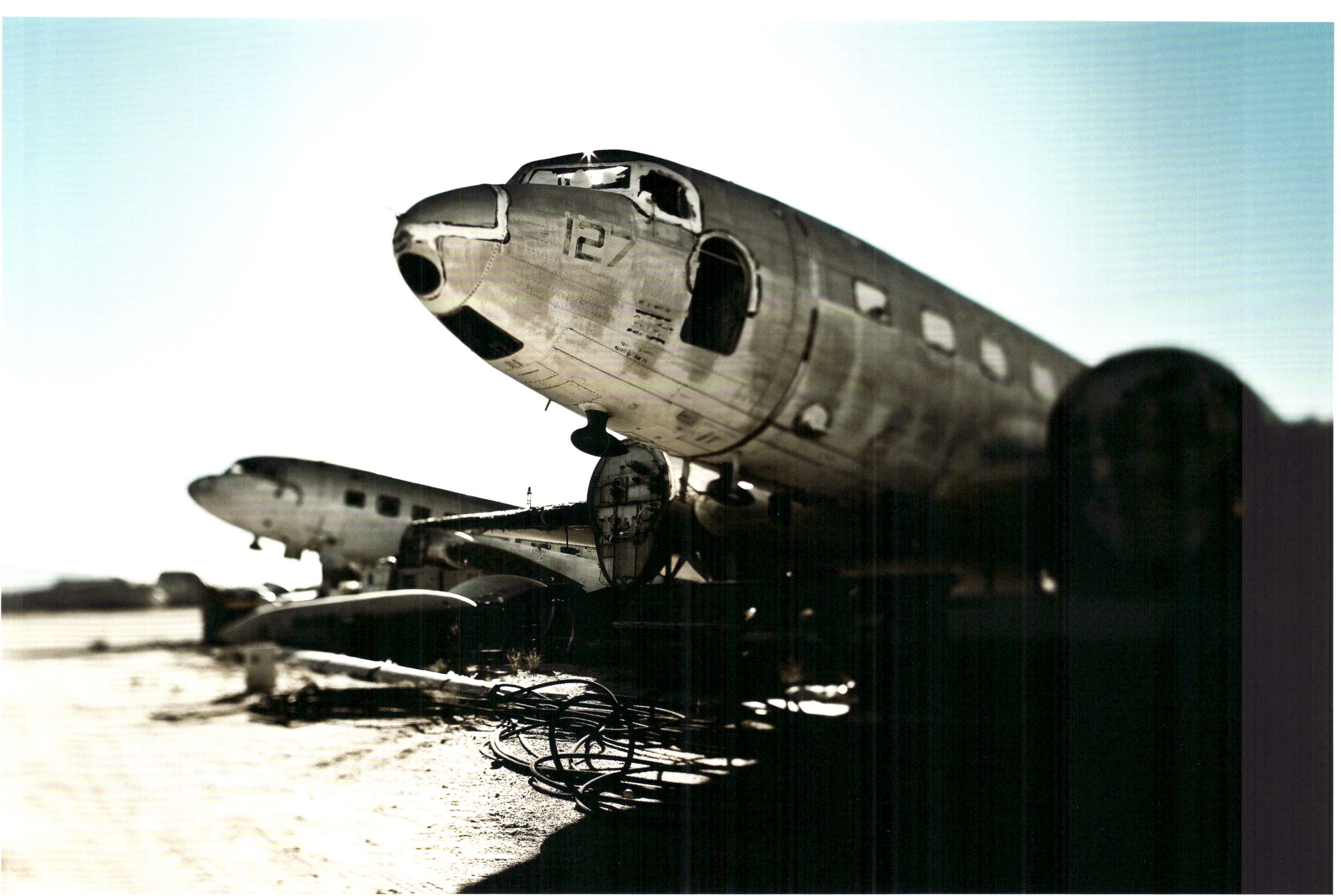
127

Skymaster #3

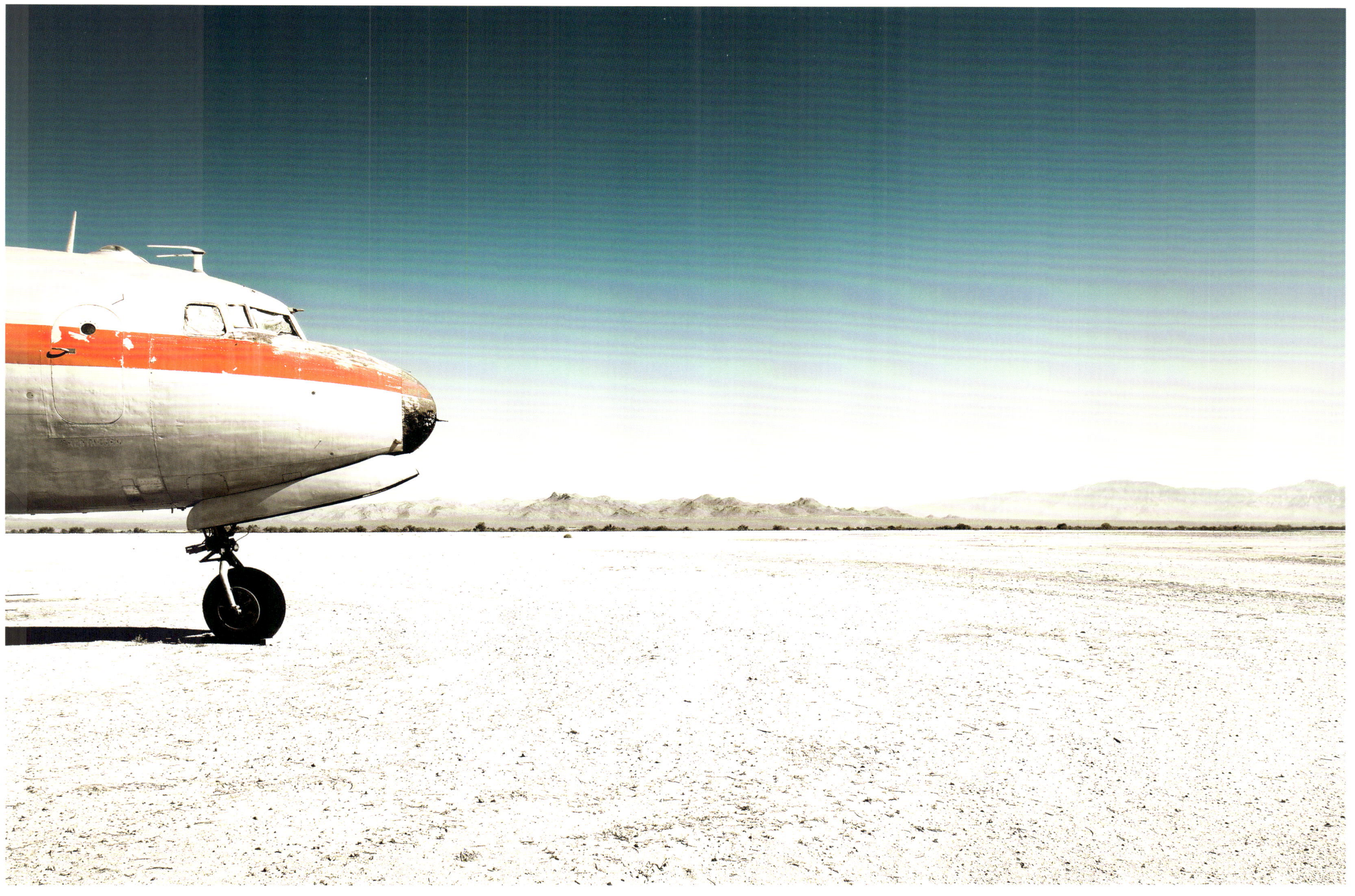

Skymaster #4

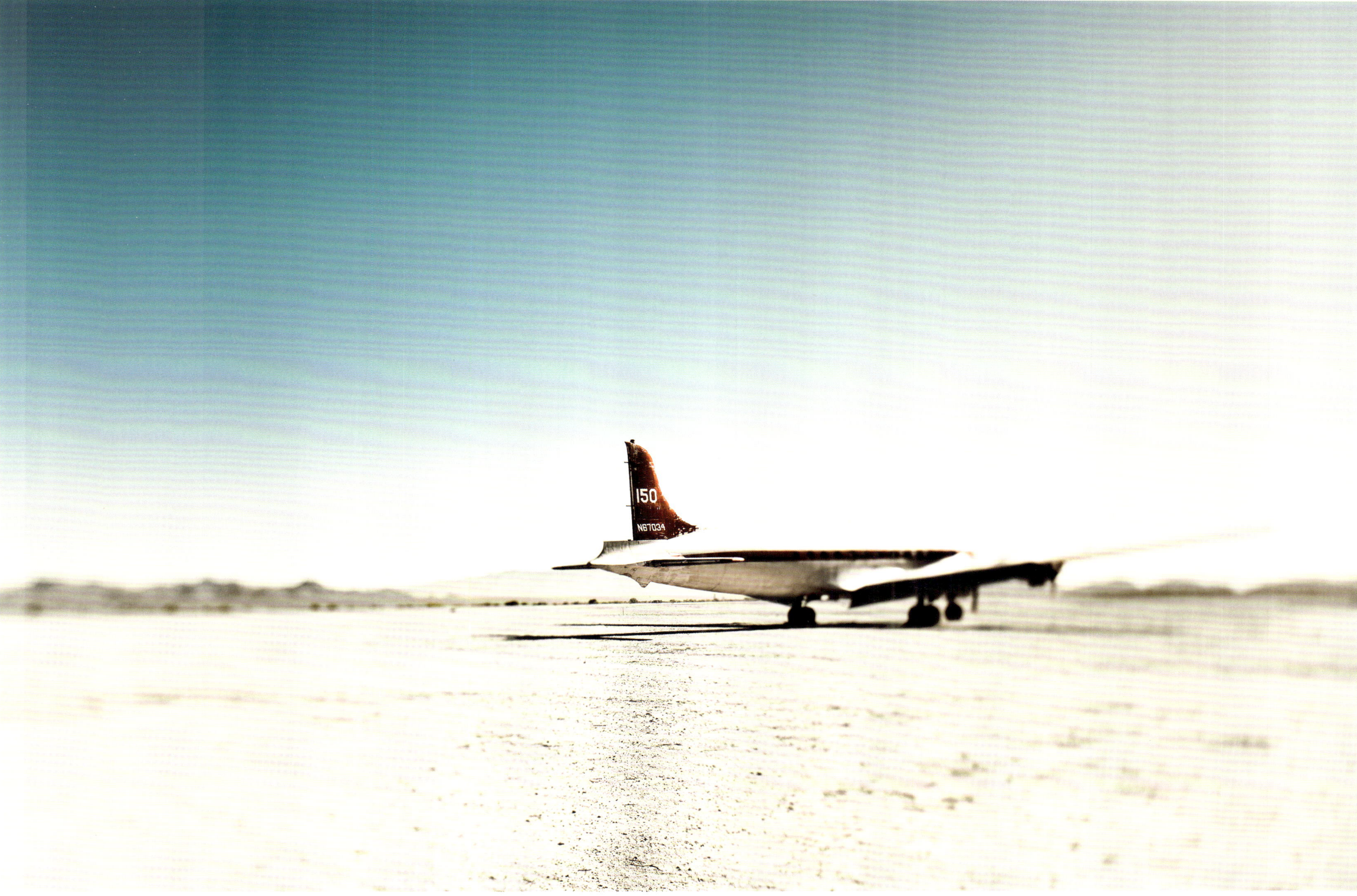
150
N67034

No Smoking

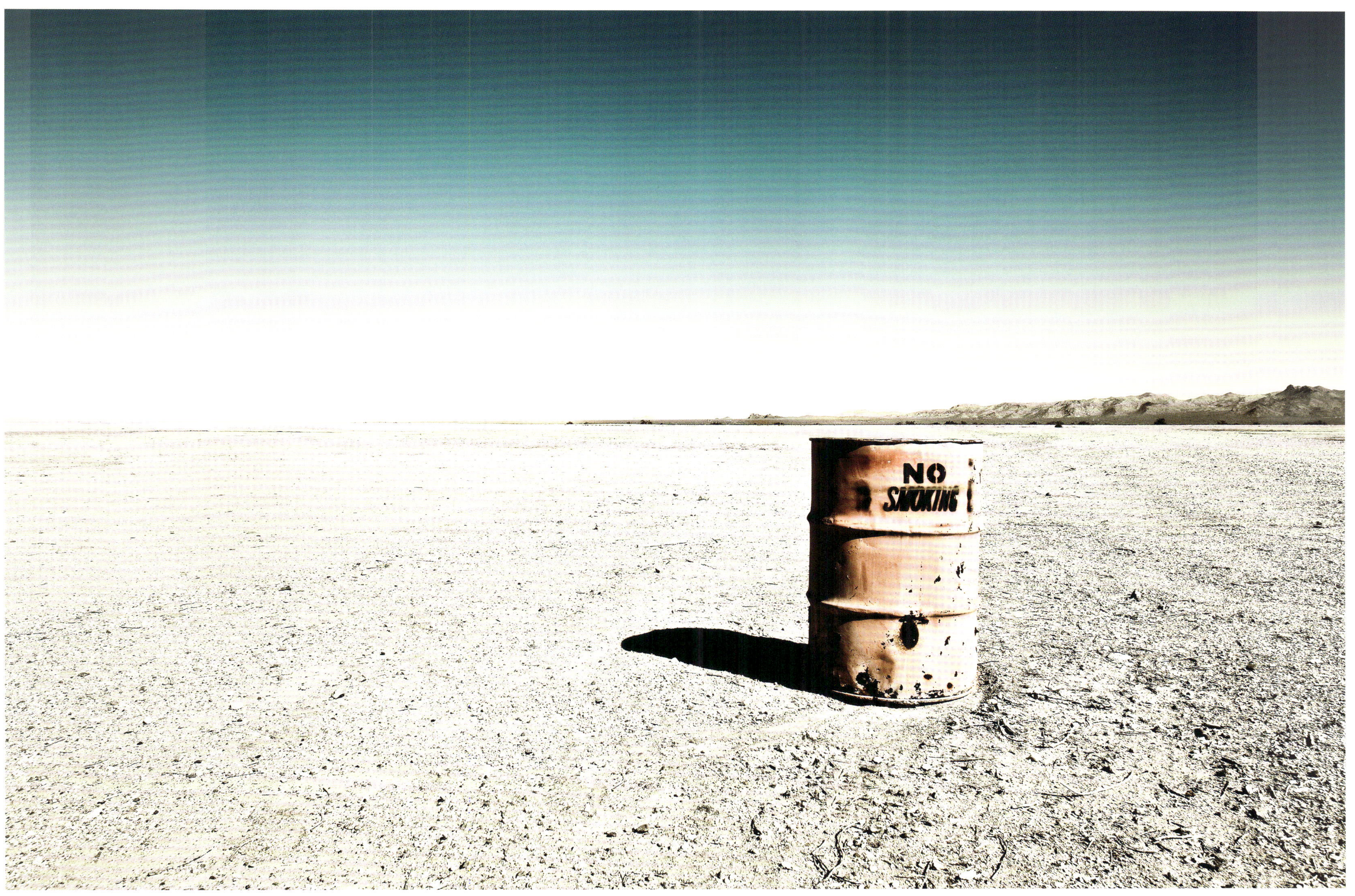

NO
SMOKING

Air Force One

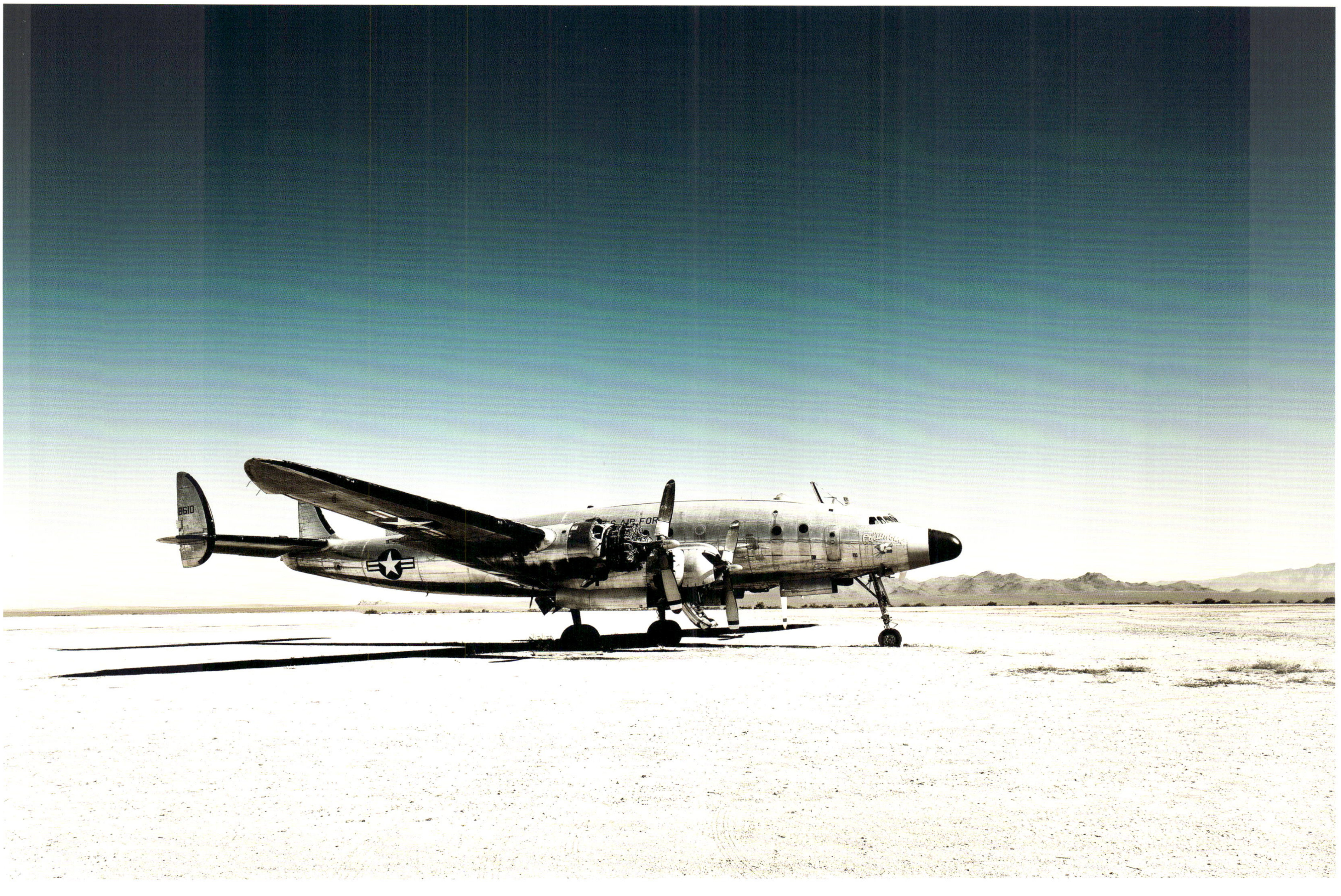

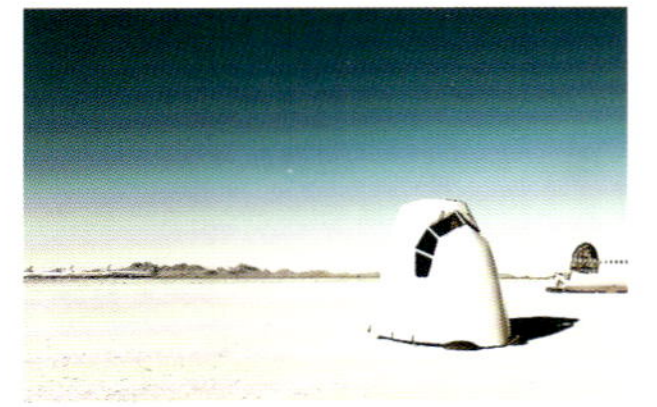

010 Cockpit

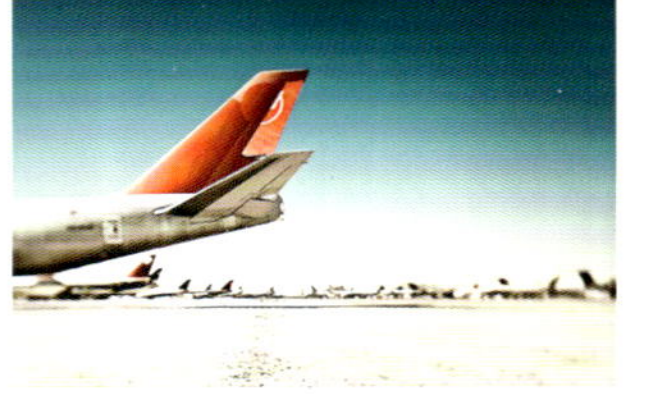

012 Tail #1

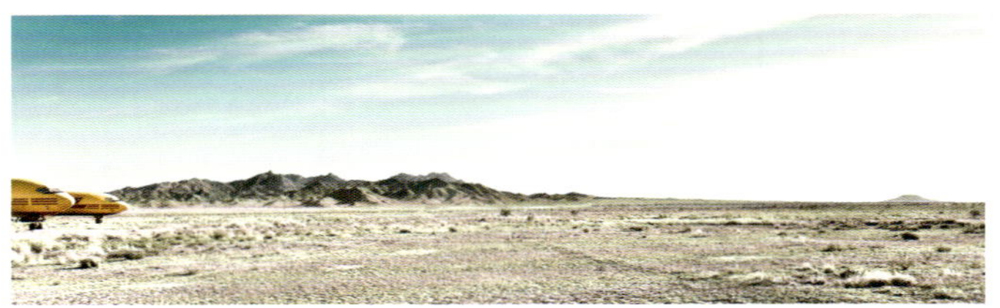

014 Stranded #1

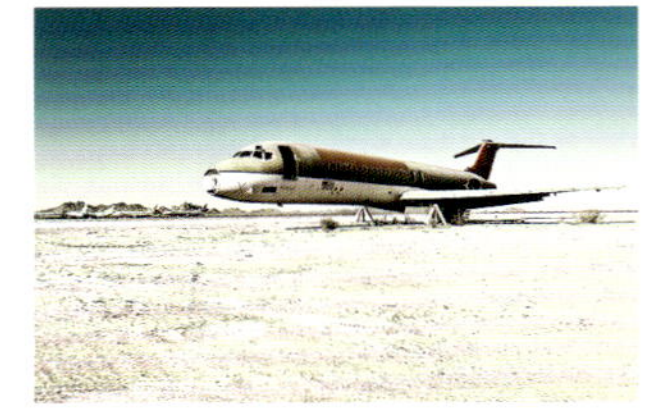

016 Fuselage #1

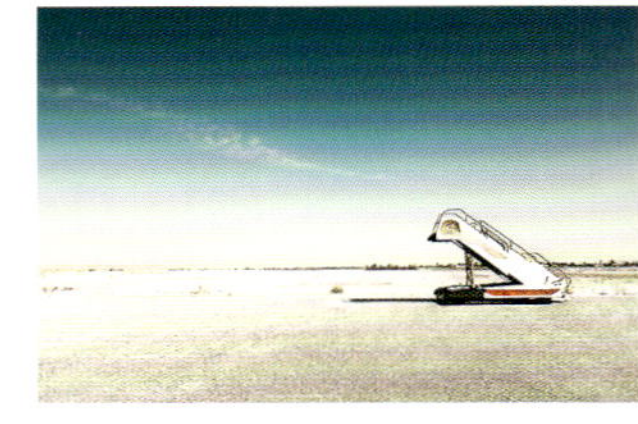

018 Gangway

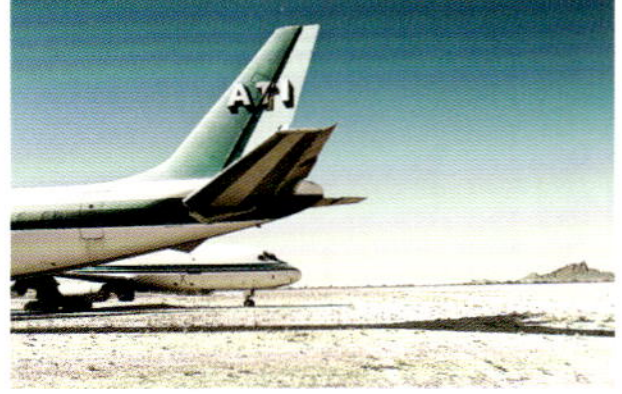

020 Tail #2

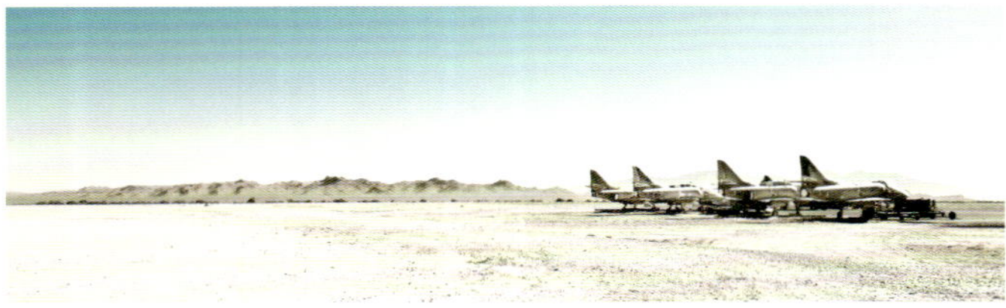

022 Stranded #2

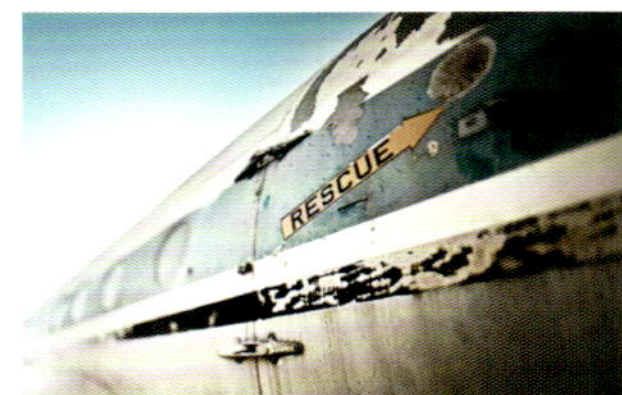

024 Rescue

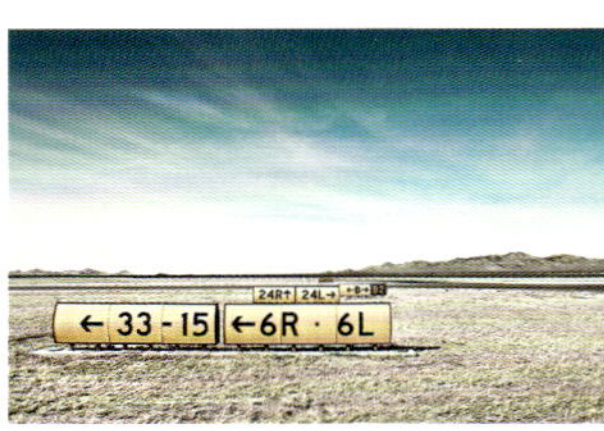

026 Taxiway

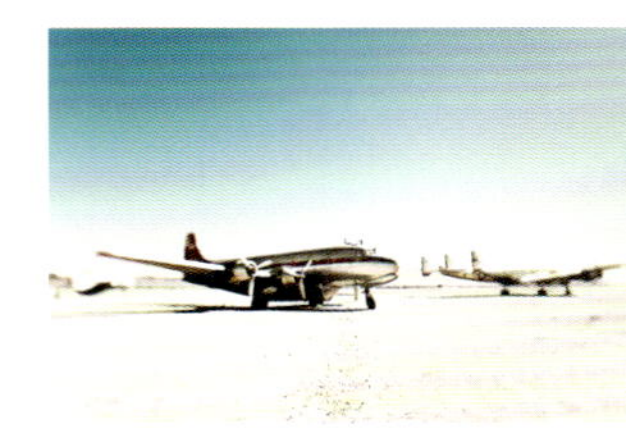

028 Skymaster #1

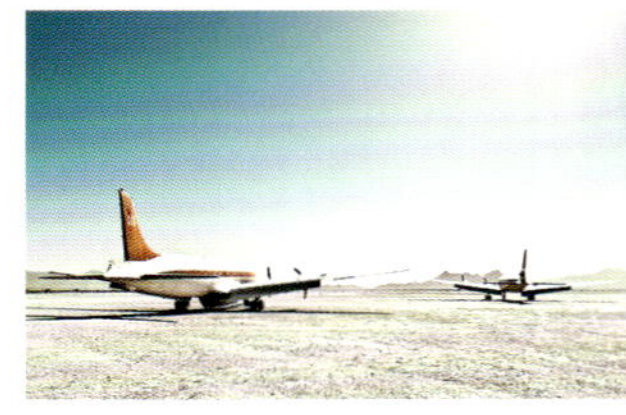

030 Skymaster #2

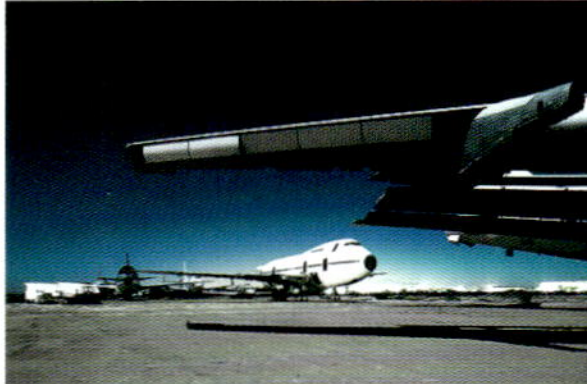

032 Full Moon #1

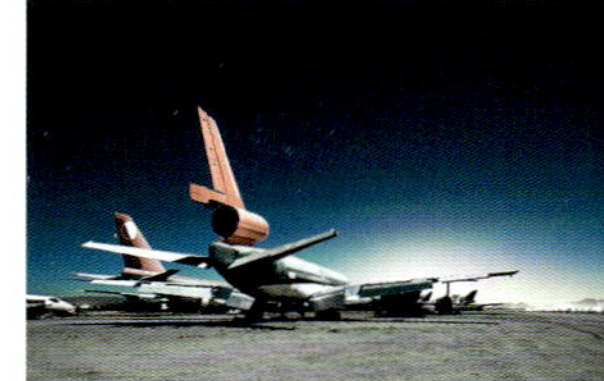

034 Full Moon #2

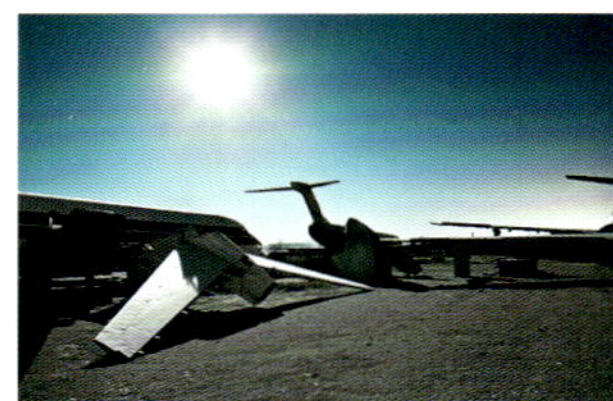

036 Full Moon #3

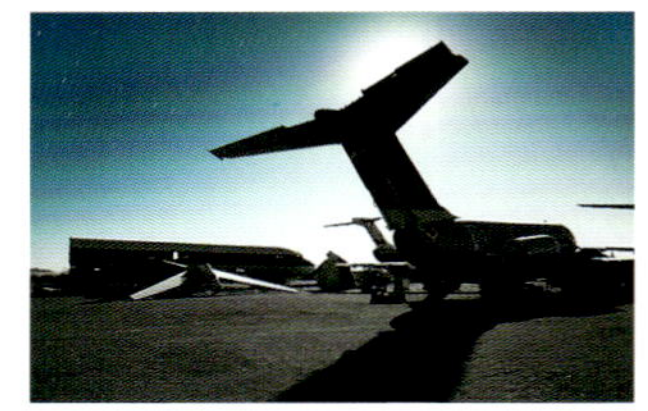

038 Full Moon #4

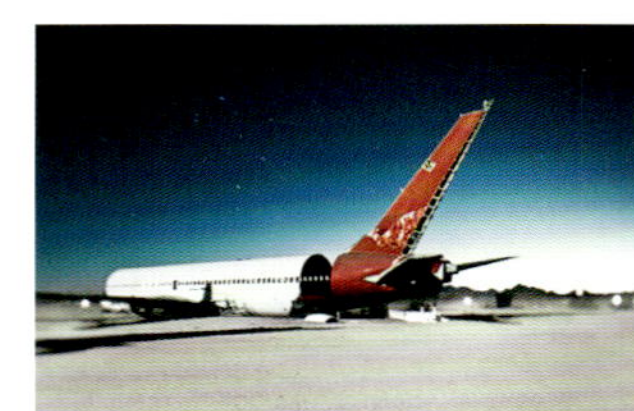

040 Full Moon #5

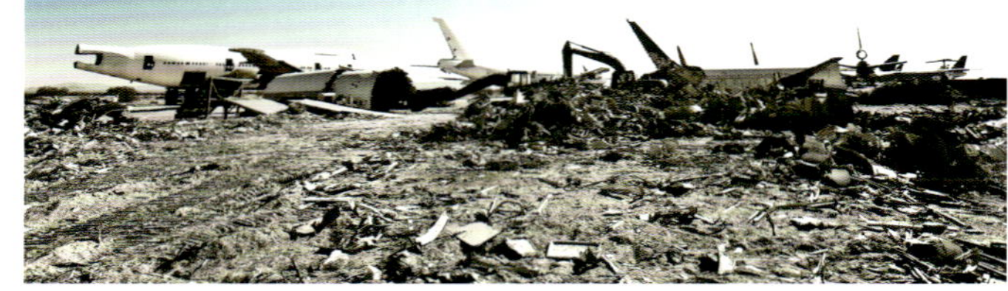

042 Boneyard #1

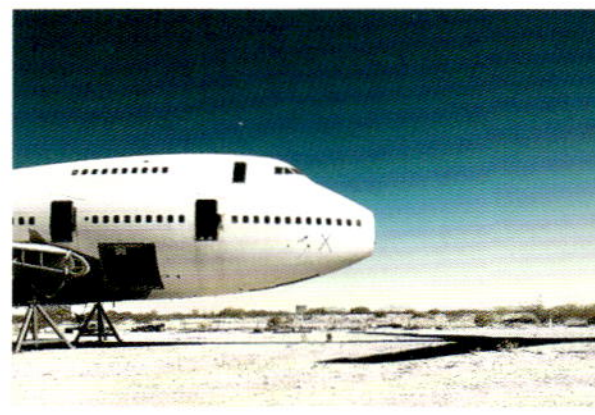

044 747 #1

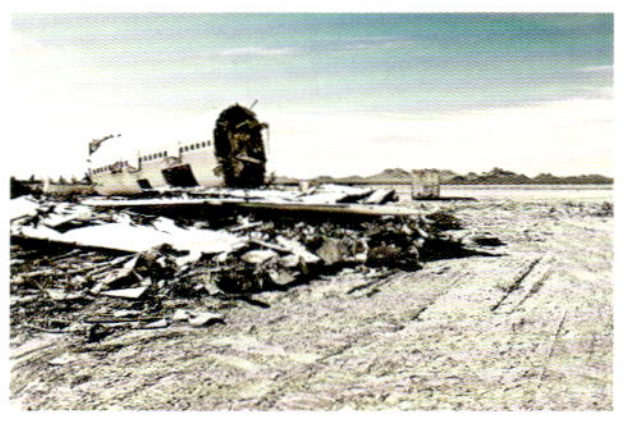

046 Boneyard #2

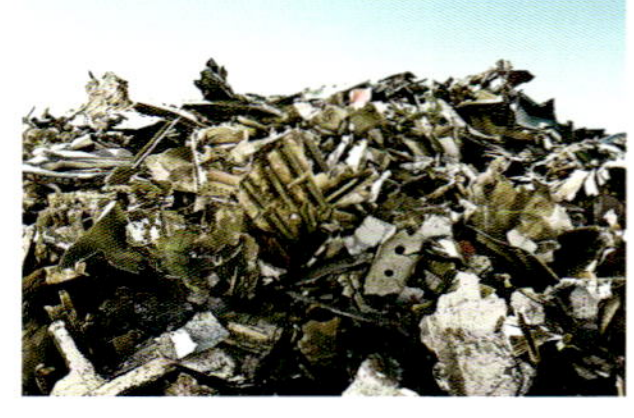

048 Scrap

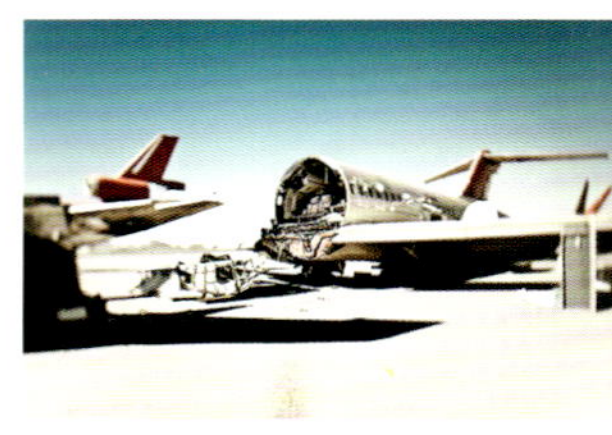

050 Clean Cut

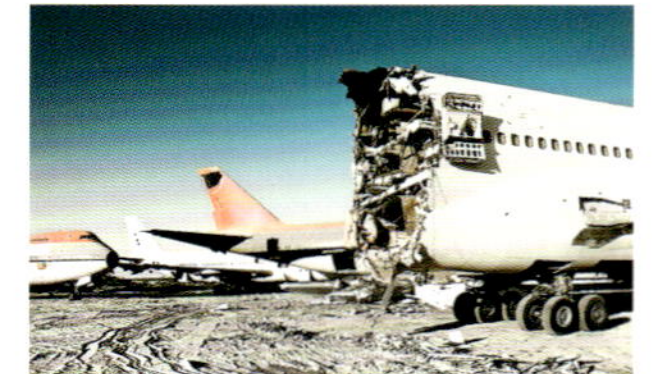

052 747 #2

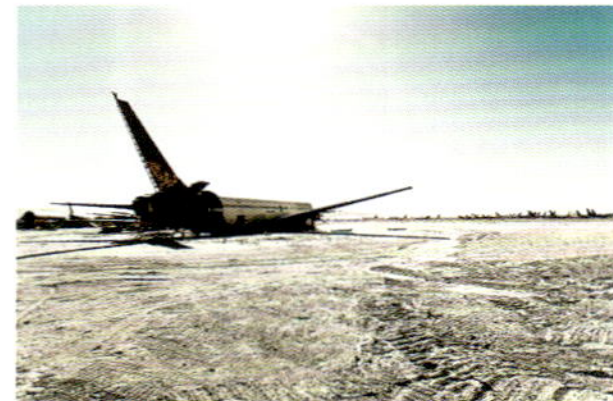

054 Grounded #1

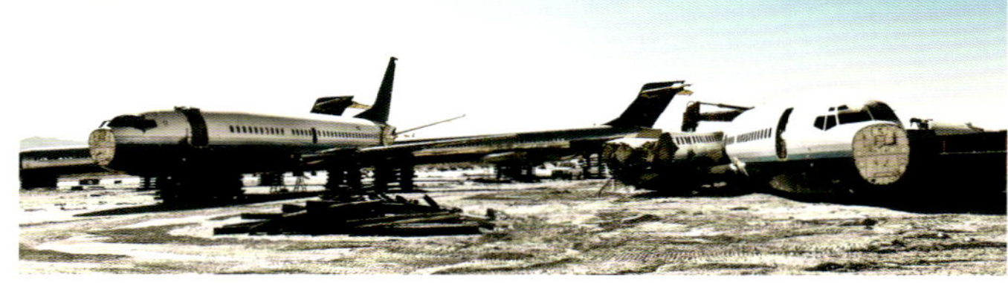

056 Boneyard #3

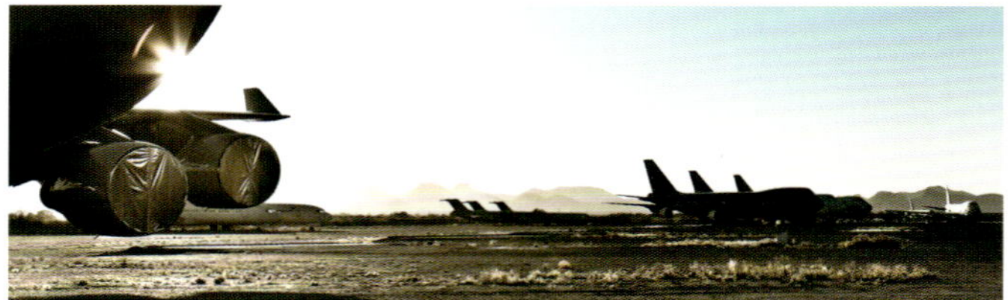

058 Stranded #3

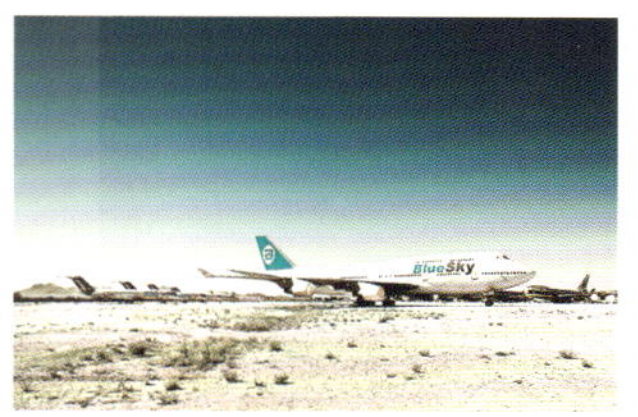

060 Blue Sky

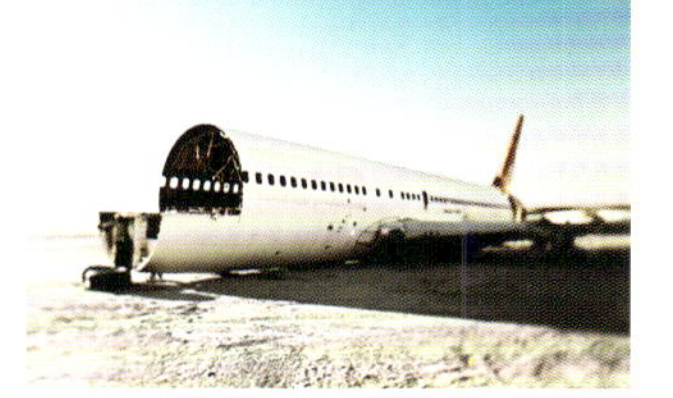

062 Fuselage #2

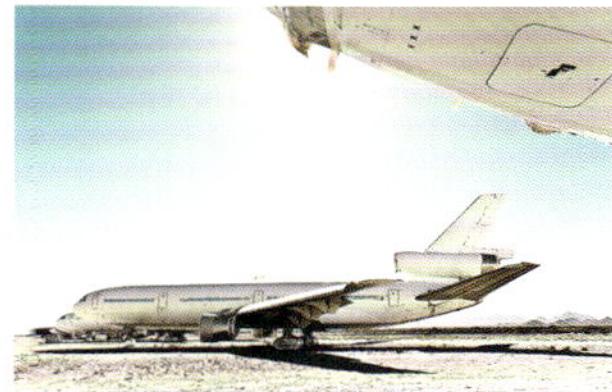

C64 DC10 #1

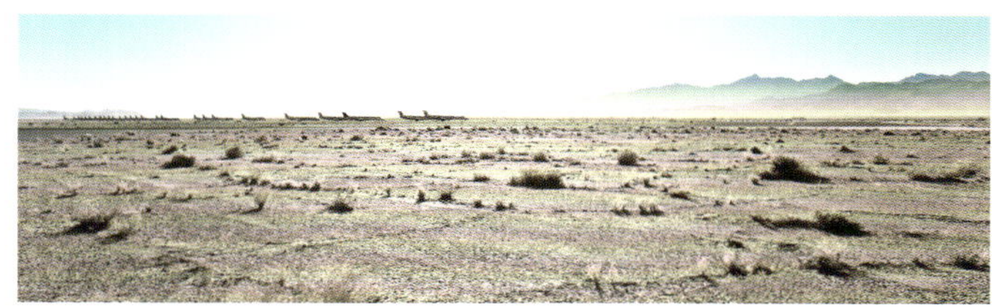

066 27 Airplanes

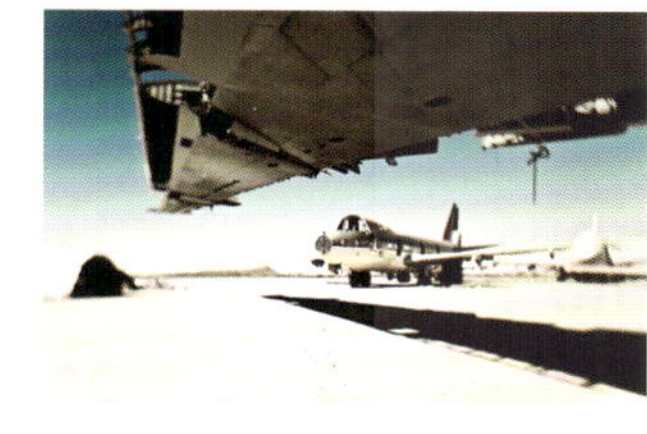

068 Grounded #2

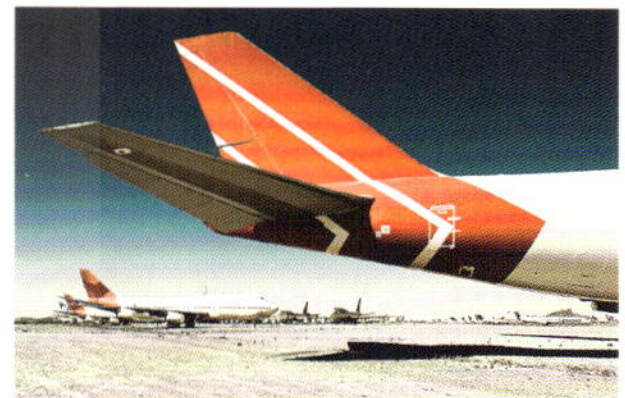

070 Tail #3

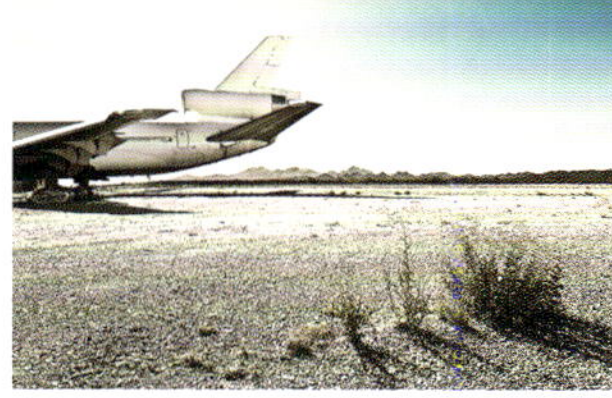

072 DC10 #2

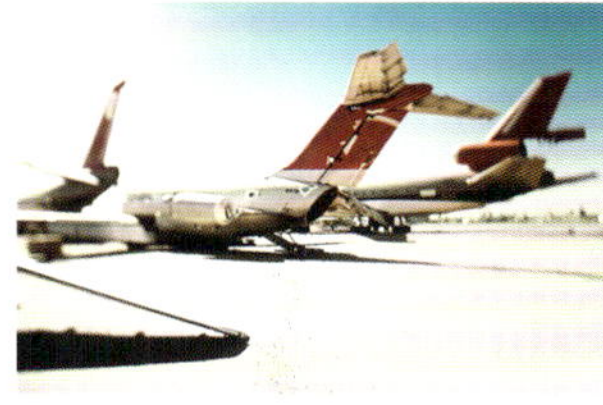

C74 Grounded #3

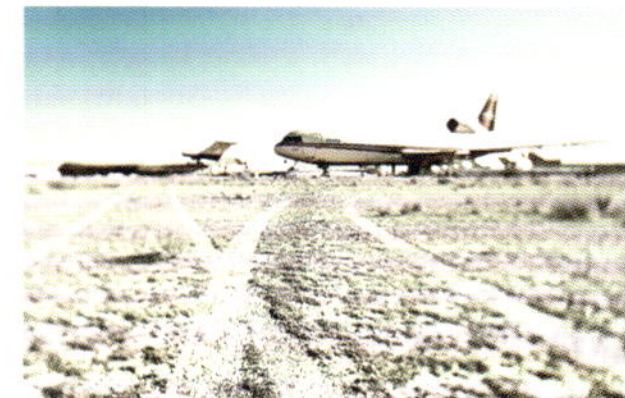

076 Tristar

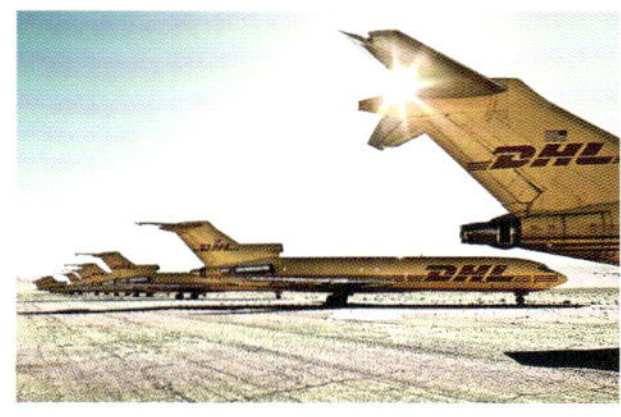

078 Grounded #4

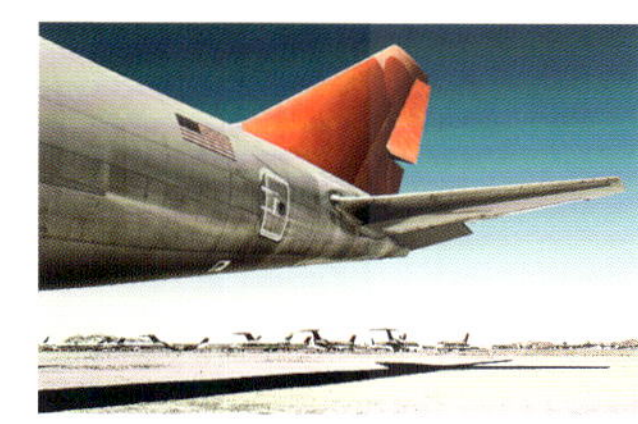

080 Tail #4

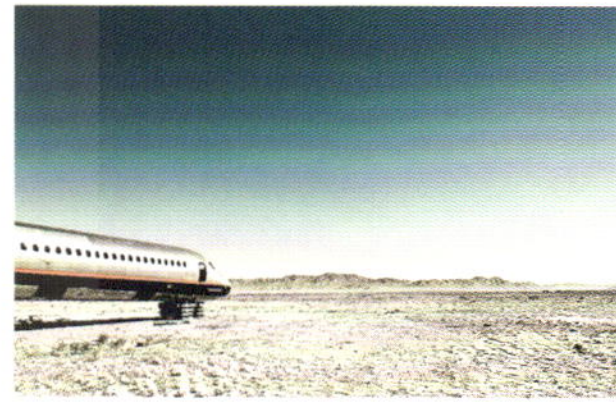

082 Stranded #4

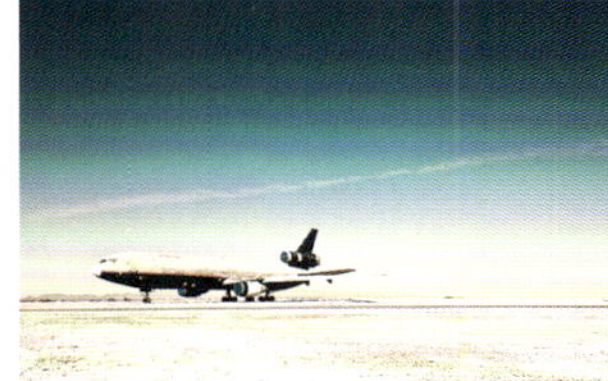

084 Freighter

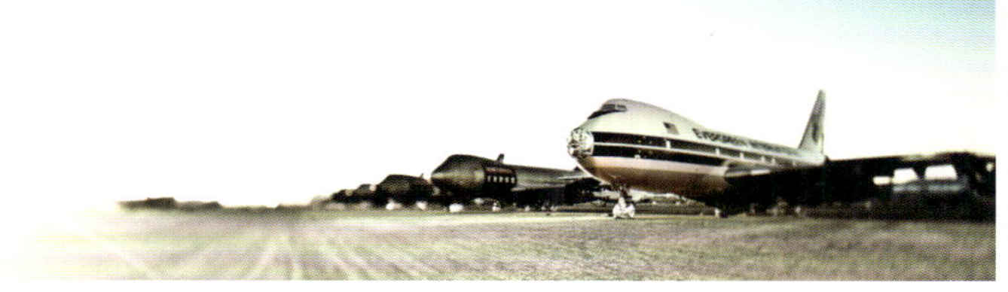

C86 747 #3

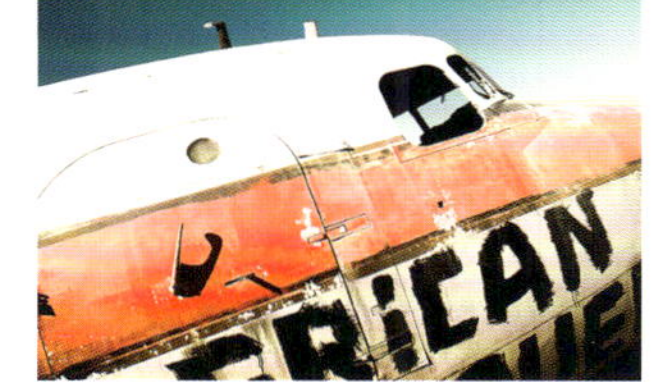

088 African Queen

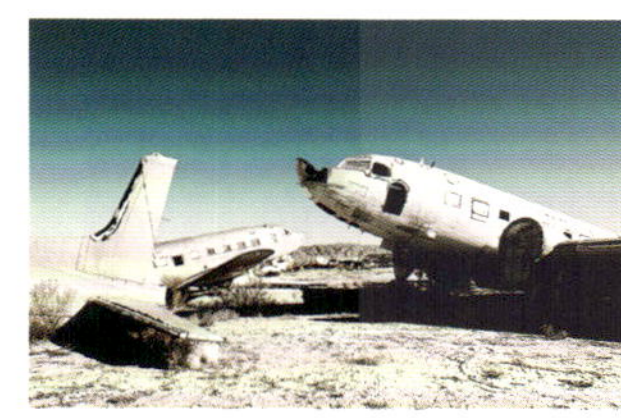

090 Dakotas #1

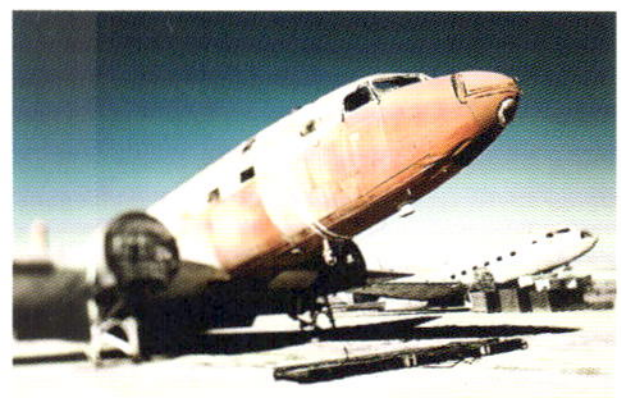

092 Dakotas #2

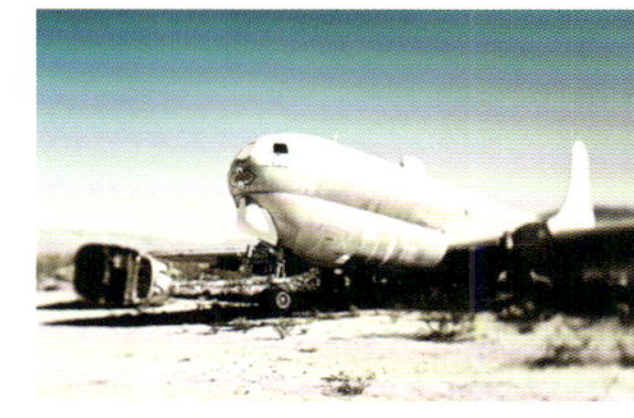

094 Stratocruiser

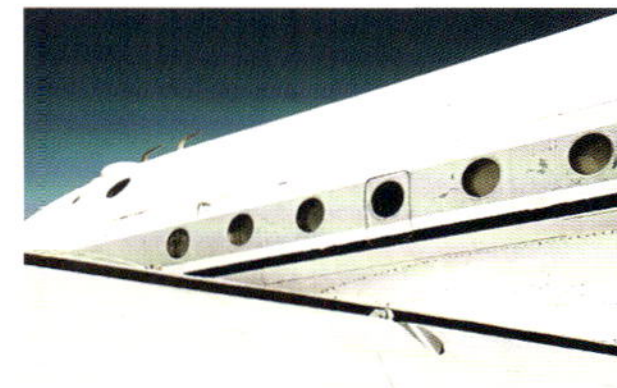

C96 Fuselage #3

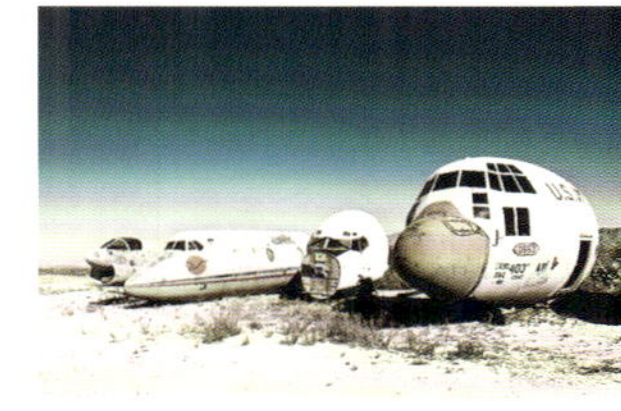

098 Friends

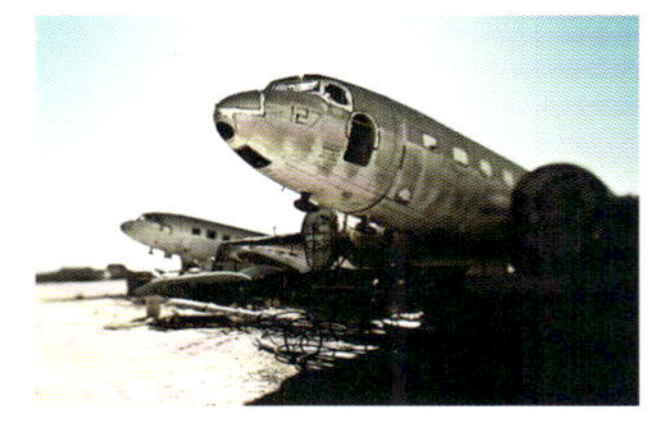

100 Dakotas #3

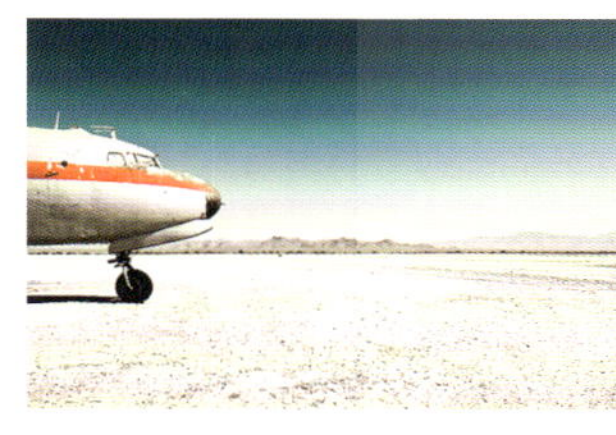

102 Skymaster #3

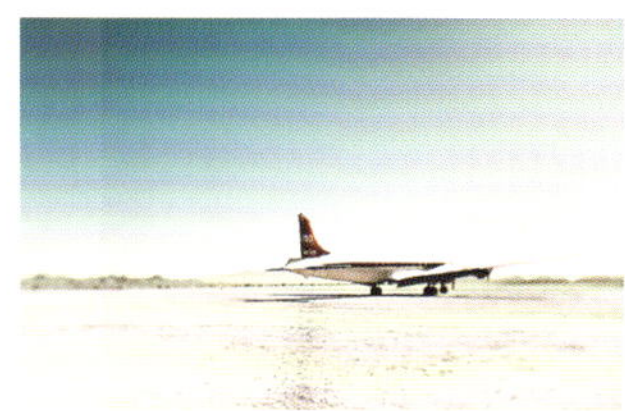

104 Skymaster #4

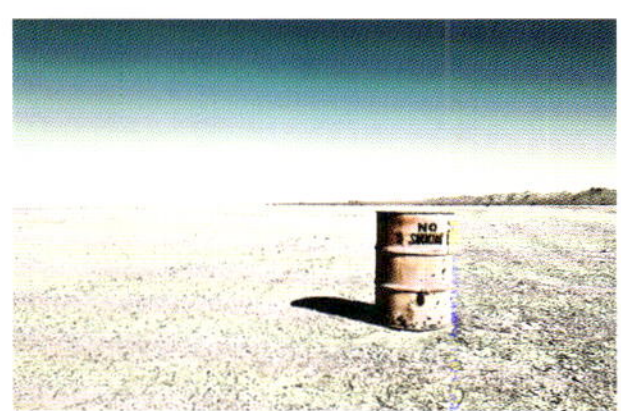

106 No Smoking

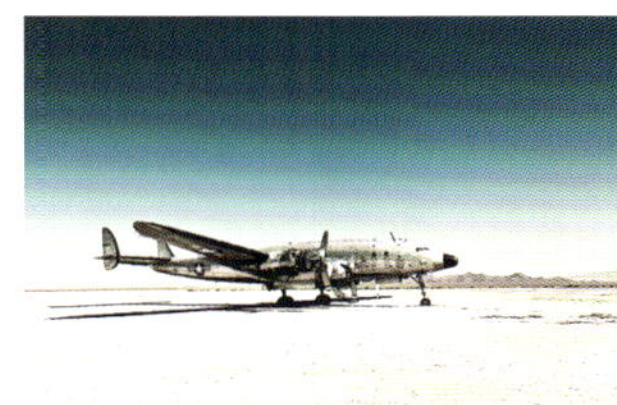

108 Air Force One

Vielen Dank an / Many Thanks to: Sonja (my heart and soul) for helping me stay organized & keep the project running, Alexander for his patience & for being the best host in the US, Bob for showing me the Hualapai Mountains at night and my first skunk, Jennifer for bringing me together with 8610, Brad for the great steak we had at Lil Abner's, Terry for introducing me to Sun & Burney, Ron for letting me climb on a Skymaster wing, Dan for being my wingman during the full moon shooting, Shari for organizing the blue sky, Jim for saving me some birds, Neil for his inspiring art, Frauke for her kindness, Randy for taking me up in the air, my brother Gerhard for the hours we spent at FRA when we were kids (Tail #2 is for you!), Ellen Dietrich, Karen Fromm, Margot Klingsporn and Hans d'Orville for their support, and all the companies that kindly let me shoot on their premises.

Herausgegeben von / Published by: Werner Bartsch
Fotografien / Photographs: Werner Bartsch
Texte / Texts: Sophia Greiff, Werner Bartsch
Verlagslektorat / Publisher's proofreading: Simone Kraft
Übersetzung / Translation: Melissa Nelson
Gestaltung / Design: Werner Bartsch, Kehrer Design (Anja Aronska)
Gesamtherstellung / Production: Kehrer Design

Bibliografische Information der Deutschen Nationalbibliothek:
Die Deutsche Nationalbibliothek verzeichnet diese Publikation in der
Deutschen Nationalbibliografie; detaillierte bibliografische Daten sind im
Internet über http://dnb.d-nb.de abrufbar.
Bibliographic information published by the Deutsche Nationalbibliothek:
The Deutsche Nationalbibliothek lists this publication in the
Deutsche Nationalbibliografie; detailed bibliographic data are available in
the Internet at http://dnb.d-nb.de.

2. Auflage / 2nd Edition

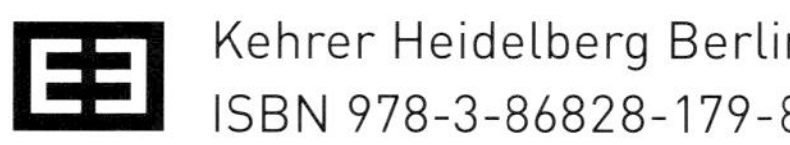

Kehrer Heidelberg Berlin
ISBN 978-3-86828-179-8